D1328486

THE EMERGENCE OF SCIENCE
IN WESTERN EUROPE

The Emergence of Science in Western Europe

Edited by

MAURICE CROSLAND

Professor of History of Science,
University of Kent at Canterbury

SCIENCE HISTORY PUBLICATIONS
New York
1976

UNIVER

First published in the UK in 1975 by
The Macmillan Press Ltd., London and Basingstoke

First published in the U.S.A. and Canada by
Science History Publications/U.S.A.
A division of
Neale Watson Academic Publications, Inc.
156 Fifth Avenue, New York 10010

© The Macmillan Press Limited 1975
© Text selection and presentation M. P. Crosland 1975
© Science in the Italian Universities in the Sixteenth
and Early Seventeenth Centuries C. B. Schmitt 1975
© Scientific Careers in Eighteenth-century France
R. Hahn 1975

All rights reserved. No part of this publication
may be reproduced or transmitted, in any form
or by any means, without permission.

ISBN 88202–041–2
LCCN 75–27320

Printed in Great Britain

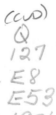
(CUD)
Q
127
.E8
E53
1976

WITHDRAWN

E. M. CUDAHY
LOYOLA
UNIVERSITY
MEMORIAL LIBRARY

CONTENTS

1. Introduction

M. P. CROSLAND
(*University of Kent*)

The seventeenth century witnessed a surge of interest in the natural world and some general historians assume, perhaps by analogy with the "Industrial Revolution" of the eighteenth century in Britain, that there was one "Scientific Revolution" in the seventeenth century, also located in Britain. But, whereas the "take off" in industry is generally agreed to have happened over a restricted period of a few decades in one country, the emergence of science was more diffuse. This was partly because of the wide area covered by what we call science, so that fundamental advances were made in different branches of science in different periods. But the spread was also a geographical one with major contributions being made in different countries at different times. The early development in Italy was not sustained and the centre of activity passed from the Mediterranean to the Atlantic seaboard, where there were important developments in France and the Netherlands as well as in England. The foundation of the Royal Society of London in 1660 is of considerable importance but the pedestrian record of this body of amateurs in the eighteenth century suggests a decline which had no parallel in the growth of the factory system in economic history. The existence of the Royal Society provided institutional continuity but hardly "take off" and to understand the emergence of modern science we must turn also to the Paris Academy of Sciences, the Ecole Polytechnique and the German universities of the nineteenth century.[1] No single movement was sufficient to produce modern science, whether the Renaissance, the Reformation, voyages of discovery, the increasing secularisation of society, the eighteenth-century Enlightenment or the French Revolution, but all contributed to change and to a situation in which people asked new questions and thought it increasingly worth while to pursue the study of the world of nature. The Italian peninsula, England, the Netherlands, Scotland, France and the German states all contributed in turn exceptional men with original ideas who helped to shape modern science.

There is, therefore, something to be said for a general review of the development of science, looking at different countries in turn. We have been fortunate in finding authors, all members of the British Society for the History of Science, who were able and willing to present their special knowledge of science from the late sixteenth to the mid-nineteenth century in countries of major importance in the development of science in western Europe. Science was one of the activities in which this part of the world made an unparalleled contribution to modern civilisation. It is clear that any choice is necessarily selective but most of the countries of western Europe which have contributed significantly to science in the period 1600–1850 are included.

The organisation of the book is around a number of generally recognised focal points of significant advance in science with a certain amount of comparison implicit or explicit in most of the chapters. Of course the first achievement was the emergence of an activity which can be fairly described as science. Several ancient civilisations and notably the Greeks, made valuable contributions towards our understanding of the natural world. A belief in the order and rationality of nature and the dignity of man under a single transcendent God, arising out of the Judaeo–Christian tradition, helped prepare the ground but did not produce science. Eventually, in the seventeenth century something more like modern science with an experimental methodology began to emerge. As the period surveyed in this book extends over several centuries it may be worth emphasising that the "science" at the beginning of our period was a much broader and less well-defined activity than the laboratory discipline called "science" in the nineteenth century.

In the cases of four countries of major importance in the scientific endeavour: Italy, England, France and Germany, contributors present different, and sometimes complementary, aspects of scientific interests and development in these respective countries. Alex Keller concentrates on technology, whereas Charles Schmitt is more concerned with the theoretical dimension and the academic context in which a special kind of learning, now called "science" emerged. Marie Boas Hall presents science at the Royal Society as an activity essentially based on experiment, whereas Piyo Rattansi explores the religious dimensions of natural philosophy in seventeenth-century England. From the standpoint of the historian of the Paris Academy of Science Roger Hahn presents a general view of the activities of men of science under the *ancien régime*, while Maurice Crosland explores some of the ways in which science became a profession in the late eighteenth and nineteenth centuries. Both focus on scientific careers and the institutionalisation of a career structure is analysed. The study of scientific institutions provides only one aspect of scientific activity and Wilfred Farrar's paper, which explains the importance of Liebig's laboratory at Giessen is complemented by David Knight's survey of philosophical currents in early nineteenth-century German thought. Although England, France and Germany are the obvious countries which demand an assessment, we have felt that to include smaller countries like the Netherlands and Scotland not only provides additional information about the growth of science but provides a better balance and a truer European perspective.

There is an increasing tendency to see the study of the natural world as one aspect of general culture. Although the study of the Italian Renaissance has been dominated by art historians, the relation of art to science through anatomy and architecture has long been pointed out and if Galileo did not belong to the same generation as Marcilio Ficino at least one can see a connexion through Neoplatonism and other more practical strands. Building on the political power and prestige of France in the late seventeenth century, Louis XIV was able to add the Paris Academy of Sciences as another gem in the French crown and thus gave science in his country an impetus which stood it in good stead in the eighteenth century. If the rise of science in the north of the Italian peninsula and then in France are late offshoots of a general cultural development, the rise of science in Scotland is much more closely a part of that mid-eighteenth century phenomenon, the Scottish Enlightenment.

Science is also clearly related to economic development. That the busy dockyard at Venice was the *cause* of Galileo's interest in mechanics would be extremely difficult to establish[2] but that it was a *stimulus* seems highly probable. Similarly, Lavoisier's work on nitre and oxygen can be related to his work on nitre for the government gunpowder monopoly, without committing the historian to a doctrine of stark economic determinism. It is a truism of economic history that trade and industry may flourish in response to a crisis, whether a war, a blockade or some other change in raw materials or markets. War or natural rivalry in peacetime may be incentives to the development of new technology and ultimately of scientific advance. In both sixteenth-century Italy and France at the time of the Revolution, warfare provided new incentives and opportunities which affected the course of science. In the short term war has generally had a stultifying effect on science, interrupting communications and diverting attention from the understanding of nature to more immediate problems. In the long term, however, its demonstrable utility has meant that science has attracted increasing support from governments and has developed from a minority intellectual pursuit to a major activity involving national policy.

In the past, blatant nationalism has combined with myopic parochialism to prevent a balanced view of scientific achievements in other countries. A British chemist remarked in 1824 that "In France men swear by Lavoisier, Berthollet and Gay-Lussac; in Germany by Stahl, Richter and Stromeyer; while in England they appeal to Black, Priestley and Davy."[3] In other words, one tends to have an exaggerated opinion of the importance of one's fellow-countrymen. The justification for the existence of this book is that in the emergence of modern science from the late sixteenth century to the nineteenth century no country has held a monopoly.

Medieval Europe, which gave rise to the universities, was characterised by a basic religious unity as well as the common use of Latin as the international language of learning. In the sixteenth century the face of Europe was scarred by wars of religion with various political implications. But whatever religious or civil authority was to be accepted, man could unite in the humble study

A*

of God's creation. (Nor after the Renaissance was man necessarily so humble.) In the seventeenth century there was a conscious effort to present science as a universal enquiry, above politics and religion, and in the new scientific societies these contentious and divisive areas of discussion were specifically excluded. Correspondence on an international scale was conducted by those universal intelligencers, Mersenne, Hartlib and Oldenburg. With his many foreign contacts, Oldenburg did much to make the Royal Society more than just an English Society. A century later Volta addressed his major memoir on electricity from Como to the President of the Royal Society in London[4] and also travelled to Paris to present his work to the French National Institute. Such correspondence and travel helped the cause of international science and in this way science developed as an aspect of that republic of letters which transcends national boundaries. Expeditions to other parts of the world for astronomical or geodesic purposes helped to encourage international co-operation and compensated for the growth of nationalism in the political sphere in the nineteenth century. Vast improvements in communication have helped in the twentieth century to erode national differences in science and to give it some general homogeneity.

There were, of course, other undercurrents and mercantilist competitiveness might well have overflowed into the intellectual realm. National pride also influenced the evaluation of scientific achievement and Descartes continued to have a large following in France when the Royal Society of London accepted the leadership of Newton. Reverence for the work of Newton reached an absurd level later in the eighteenth century, when British mathematicians followed their compatriot to the exclusion of important continental developments by the Bernoullis, Euler and Lagrange.

Perhaps we see nationalism at its worst in the nineteenth century. During the Napoleonic wars we find national rivalry reflected occasionally in the work of scientists and Humphry Davy prepared a lecture in which he spoke of "the scientific glory of a country" and declared that there was "one spirit of enterprise, vigour and conquest in science, arts and arms".[5] But there was no stronger rivalry in western Europe in the nineteenth century than that between France and Germany. Liebig reacted against the claims for French chemistry; he said that Lavoisier and his colleagues by introducing a new nomenclature had "blotted out" the achievements of the phlogistic chemistry of Becher and Stahl so that "To many the knowledge we now possess appears to be only the inheritance of the French school of that day".[6] Liebig was therefore concerned to present a different perspective of chemistry emphasising, for example, Richter's law of equivalents and saying: "The discovery of that law must be attributed to the sagacity and acuteness of a German chemist and the name Richter will remain as imperishable as the science itself".[7]

At the beginning of our period we are far removed from nineteenth-century nationalism but there was a linguistic factor which tended to reinforce any geographical differences. The rise of the different vernaculars to replace the Latin of the medieval Church and universities had its effect on science as well as on literature. Descartes wrote his *Discours* in French, the Dutch mathematician Stevin was an enthusiastic proponent of the vernacular, Galileo made a point of writing his famous dialogues in Italian and Boyle composed his numerous tracts in English. By the eighteenth century Latin had largely disappeared from scientific literature, although the Scandinavian countries[8] with minority languages had an obvious motive for maintaining what had been the traditional international language of learning. The use of the vernacular gives further justification to the use of such descriptions as, for example, "French science" but it is obvious that a language is a reinforcement of an existing cultural situation rather than a controlling factor. The most important effect of the use of the vernacular from the point of view of this book was that it tended to cut off one country from another, so that men of science in neighbouring countries like France and Germany could develop their own ideas often in ignorance of the work of their neighbours.

One of the dangers about speaking of science in terms of country is that the unwary reader might assume that scientific interests and expertise were distributed homogeneously throughout the land. Obviously, whatever the country, cities and universities tend to be focal points of intellectual activity. Dr Keller makes it clear that his "Italian science" refers to activities in certain states in Northern Italy like Tuscany and the Venetian republic. France is a large country by the standards of western Europe but increasing centralisation since the seventeenth century has meant that national aspirations have been increasingly concentrated in Paris. Nevertheless, the academies founded in provincial towns in the eighteenth century give some indication of intellectual activity outside Paris. The Academies of Bordeaux, Dijon and Lyons were among those making contributions to science and Montpellier maintained some of its ancient medical tradition. The foundation of the National Institute in Paris in 1795 had the unintended effect of once more focusing activity in the French capital.

From other parts of the world it is often assumed that Great Britain or the United Kingdom forms a convenient unit so that one can always speak of "British science". This may be a useful description today but it was certainly not so in the eighteenth century when, despite and perhaps because of the Act of (political) Union of 1707, Scottish culture asserted itself in a distinctive way. But even in England one cannot always use the phrase "English science". In no country of comparable size have there probably been greater local differences in the scale and type of scientific activity. The fact that the industrial revolution of the eighteenth century was concentrated in the midlands and the north, far from such fashionable centres of culture as London and Bath, meant that different social and intellectual activities came to flourish in these areas. The Manchester Literary and Philosophical Society (founded 1781) was one of the principal provincial groups which contributed to the growth of science. Fortunately provincial science in this period has been well studied recently[9] and it is not necessary here to re-emphasise this activity. At this time the two ancient universities made little contribution to science but by 1830, when Charles Babbage was writing on the decline of British science,[10] there were already signs of a revival in Cambridge. Babbage himself together with John Herschel, George Biddell Airy and others contributed to this. The 1830s were unparalleled in discussion of reform in science as well as politics and with the foundation of the British Association for the Advancement of Science (1831), combining English and Scottish elements, science found a new focus.

A geographical approach to the origins of famous men of science was advocated by H. T. Pledge when he called attention to the clustering of the birthplaces of famous scientists.[11] A prominent feature of his general text-book of the history of science was a series of six maps of different parts of Europe, indicating places of birth. After a cursory examination of the contribution of Ireland[12] in the nineteenth century, he rejected his own preliminary hypothesis of a simple correlation between numbers of scientists and wealth and population. If the rise of science can be expressed by a formula, it is certainly not a simple one.

In technology similarities and differences are sometimes explicable in simple geographical terms. Historians of technology sometimes speak of "alpine technology"[13] and the similarity between problems of drainage faced in Holland and the English Fenland led to the employment of similar methods and even the same personnel.[14] National conditions and interests help to explain why in early seventeenth century London the professors of Gresham College took more than a passing interest in navigation.[15] In central Europe in the eighteenth century there were no problems of navigation but there was considerable mineral wealth which provided adequate justification for the establishment of mining academies such as those at Freiburg (1765) and Schemnitz (1770).[16]

In the late nineteenth century the botanist Alphonse de Candolle undertook a study of the distribution of scientific genius among different populations. Although many of his arguments are naive and must be seen against the background of nineteenth century educational and biological debates on heredity and environment, his book does have some bearing on the emergence of science in different national contexts.[17] He felt that religious and political freedom were basic conditions for science to flourish but was unable to explain how science blossomed in France in the seventeenth and eighteenth centuries under conditions allowing a minimum of freedom. As a Swiss Protestant, Candolle took pride in pointing out what important contributions small Protestant countries such as his own had played in science. He asked what influences would create an environment favourable to the growth of science and considered that here religion was more relevant than nationality. In this century Robert Merton has extended the idea of a correlation between contributions to science and Christian denomination but without developing its implications for national influence.[18] Merton's ideas of a simple correlation between the rise of science and "Puritanism" have recently been criticised by several leading historians of science.[19] The trouble about any general thesis which explains the emergence of science in relation to one specific factor is that a good case often depends on the selective use of evidence and, while the thesis may be plausible in one area, it fails in another.

The contribution of the Huguenots to science is seen in better perspective when one has studied evidence in England of recusant science under Elizabeth I and the achievements of the Dissenters at the time of George III (not to mention the work of Jewish scientists in the twentieth century). One is then less impressed by the achievements of any one religious group than of the role of minorities, which managed to overcome barriers to gain access to some means of higher education. Often they turned to science as an alternative to political power and important social positions from which they were barred. The case of Scotland's reaction to its political emasculation in the eighteenth century may seem exceptional until one has considered the Prussian reaction to military defeat by Napoleon I and the foundation of the University of Berlin in 1810. Thus defeat in one field may be compensated on a national scale by intensified intellectual endeavour. In the eighteenth century individuals and minority groups, deprived of basic political rights, looked elsewhere for a means of self expression. This is one reason why the Dissenters in England and the middle classes in France were able to make important contributions to science.

There are other causes where a generalisation based on limited evidence may be deceptive. Anyone who has studied science in the eighteenth century in England or France may consider universities as peripheral to the development of science and it is useful to have the temporary weakness of some universities placed in a broader context. In Italy around 1600 and again, in a different way, in Germany around 1850 universities were focal points for the study of science. A strong centralised state can be a powerful patron of science, as we see from eighteenth-century France, but the cases of the Netherlands, England and the German states serve to remind us of the advantages of free local development and even competition between local centres. From a general survey there therefore emerges a rather different picture from that which one might have after looking at only one country or one period. One is more cautious in trying to formulate general conclusions but a few definite patterns have emerged.

The idea that scientific achievement can be assigned to particular countries rather than individuals is not one which will be immediately accepted by everyone. Chekov wrote that "there is no national science just as there is no national multiplication table"[20] and certainly mathematics would seem to be an activity which is least easily related to any national background. On the other hand the flora and fauna studied by a botanist may be dependent on geographical factors. But it is not only the natural environment but social, religious and political atmosphere and education which tend to impose certain norms which may be different in different countries.

The time factor in the development of science has been appreciated much more than a corresponding place (or "cultural") factor. Ideas unknown or unacceptable in one age become current in a later period, and this is so obvious, that people sometimes forget the differences between countries (even neighbouring countries) in the same period. We may illustrate such differences by contrasting some of the characteristics of science in Britain with science elsewhere. Thus in Britain, but not in France, we find the persistence of a strong tradition of natural theology which owed much to the seventeenth-century contributions of Robert Boyle and Isaac Newton. It persisted as a main stream of thought in Britain well into the nineteenth century and in the *Bridgewater Treatises* leading British men of science including William Whewell, William Buckland, William Prout and Charles Bell demonstrated divine providence in branches of science ranging from astronomy and geology to chemistry and physiology.[21] One cannot imagine a collection of *Bridgewater Treatises* written by leading nineteenth century French scientists.

Another British characteristic which may have affected the approach to science arose out of the traditional British sentiment of kindness to animals. The experimental method of the French physiologist Francois Magendie, based on the continuous practice of vivisection, was no doubt considered repulsive by many of his fellow countrymen but it would have led to a public outcry in Britain. Yet Magendie shares with his contemporary Charles Bell credit for the "Bell-Magendie rule" (1822).[22] The apparent insensitivity of the Frenchman to the pain of his experimental animals contrasts markedly with "the distaste which Bell felt for experimental work in conscious animals".[23] Bell explained that he used experiments not as a means of discovery but to justify his conclusions to others and, in referring to work done in France, he spoke critically of "experiments without number or mercy".[24]

The importance of practical observation and experiment was widely recognised from the seventeenth century, but a major contribution to this empirical approach came from English natural philosophers. A strong element of empiricism may be seen in the work of the fourteenth-century English scholastic William of Ockham long before the foundation of the Royal Society and with the philosopher John Locke it was strengthened and perpetuated as a powerful "English influence" on the continent in the eighteenth century. A tradition of British empiricism is seen in the work of the eighteenth-century pneumatic chemists: Hales, Black, Cavendish and Priestley. In the 1770s and

1780s, Lavoisier was able to build on this edifice to produce a new general theory of chemistry. This apparent ability of the French mind to synthesise and provide a theoretical framework on the basis of empirical work done in Britain was repeated when Sadi Carnot went beyond the steam engine of James Watt to the ideal engine in his *Reflections on the motive power of fire* (1824).

The insistence of Descartes on clarity of thought and expression has been a powerful force in education in France and one understands how difficult it was for the majority of educated Frenchmen to have any sympathy with vague general philosophies of nature emanating from Germany. On the other hand the specialisation which characterised French science in the early nineteenth century and the suspicion with which all-embracing philosophies of nature were viewed led to a situation in which French scientists made few contributions to the generalisation of ideas on force and energy which led to the historic law of conservation of energy. In Germany *Naturphilosphie* provided an intellectual climate favourable to the broad approach and in Britain physics was still "natural philosophy", embracing the general study of nature and overlapping with a native tradition of natural theology. It was significantly from German and British men of science that the leading ideas of conservation of energy came. The simultaneous discovery of conservation of energy in the 1840s provides some support for an idea of national characteristics in science in so far as the Englishman Joule based his conclusions solidly on experiments whereas the German work was more obviously connected with a metaphysics.[25] Although the conclusions arrived at by C. F. Mohr resembled in some ways the work of Faraday and Grove, Mohr's philosophical inspiration and sudden leaps in argument contrast markedly with the sober British account of conversion processes based strictly on experimental researches.

The concept of a "French genius" distinct from an "English genius" was presented by the distinguished French historian of science Pierre Duhem, who argued the case for national styles in physics.[26] Duhem observed British physicists continually having recourse to mechanical models and wondered whether these had been inspired by the mills and factories of Victorian Britain. Duhem writes "We thought we were entering the tranquil and neatly ordered abode of reason but we found ourselves in a factory". If the model was not normally a crude one of cogwheels and pulleys it might be marbles or billiard balls to represent atoms or pieces of elastic to represent lines of force. However, the French or German physicist (and Duhem put them together) would see a field of force in terms of a mathematical equation. The basic difference of approach was most strikingly illustrated in physics text books. It meant that some British physicists were baffled by the French abstract approach while continental physicists could not understand the qualitative talk about models. To find some explanation of the uneven use of mathematics we must consider how scientists were trained in different countries.

Education is an important factor which might contribute to producing a national culture and we would be particularly interested in any national system geared to the transmission of a common cultural heritage. France provides a better example of this than Britain, which was comparatively late in having any national educational system. In the Napoleonic University French education became established in a centralised form which lasted from the early nineteenth century to 1968. National patterns of training were imposed and the aspiring physicist, for example, had to pass rigorous examinations in mathematics at the *baccalaureat* and the *license* before passing on to post-graduate work. Of course the British physicist might also have studied mathematics at a university but it was possible to achieve recognition in physics (called "natural philosophy") without such training, as the cases of Faraday and Grove illustrate.

A glance at the educational background of leading English scientists around 1800 shows that many—if they had received an advanced education at all—had been associated with minority movements (for example the Dissenting Academies) or were educated abroad. The leading figure in optics in early nineteenth-century Britain, Thomas Young (1773–1829) had graduated in medicine at Göttingen.[27] The engineer Marc Isambard Brunel (1769–1849) was a French royalist refugee, whose son, Isambard Kingdom Brunel, one of the great engineers of all time, was brought up in England but was sent to France to learn mathematics.[28] At the University of Cambridge particular emphasis was laid on mathematics but purely as an intellectual discipline for a restricted clientele and certainly not for aspiring engineers. Individual initiative was encouraged and in new fields like science and technology there were no generally accepted courses of training.

England was remarkable for her untutored geniuses and never more so than in the middle period of the Industrial Revolution when economic and social change provided many new opportunities. William Smith (1769–1839), "Father of English geology" received only the most rudimentary formal education and, although Joseph Priestley and John Dalton were more fortunate in their schooling, which included instruction in mathematics, they were largely self taught in chemistry, to which each was to contribute so much. Of the Cornish genius, Humphry Davy, Berzelius said that if he had studied chemistry systematically in his youth he would have advanced it by a whole century instead of leaving merely a few brilliant fragments. Michael Faraday too had little formal instruction; he was a bookbinder's assistant who became Davy's assistant by chance. Among other untutored geniuses of that period we might include the Nottingham mathematician George Green (1793–1841). The necessity for self education no doubt meant that many a potential English scientist was lost but in those with the necessary talent and persistence one finds a certain independence of thought and above all originality which might have been stifled by a long period of formal training.

Much of French science has contained a prominent streak of positivism, if it is permissible to use this term to characterise some of the work of Lavoisier, more than a generation before the term was used by Auguste Comte, in his *Cours de philosophie positive* (1830–42). Lavoisier based his new chemistry on elements, defining elements in terms of what could be separated in the laboratory. He rejected atoms as "metaphysical",[29] leaving it to the Englishman John Dalton to develop the concept of the chemical atom. Lavoisier's compatriots showed strong resistance to the idea of atoms, although suspicion about accepting unobservable entities as the basis of a science was not confined to Frenchmen. However, in the history of nineteenth-century chemistry French scientists[30] figure prominently in the critique of many new and fruitful ideas including catalysis and structural formulae and the negative approach to these questions by the Frenchmen Sainte Claire-Deville and Gerhardt may be contrasted with the bolder speculations of Berzelius, Butlerov and Kekulé. It was, however, a German chemist, Kolbe (1818–84), who uttered the most strident criticism of the concept of spatial formulae;[31] he saw it as a reversion to *Naturphilosophie*, which had so dominated science in his country in his youth. Thus the concept of a national style needs to be approached with caution not only because of the many exceptions to any generalisation but because what was largely true of "German science" at the beginning of the nineteenth century had become no more than a minority movement by the end of that century. If, like the eighteenth-century Newtonians, one takes Cartesianism to stand for speculation, there may appear to be a paradox in seeing both speculative and anti-speculative attitudes emerging from the same country. Here again one must distinguish periods. The Cartesian triumph came in the seventeenth century. Leading French thinkers in the eighteenth century were deeply impressed by the experimentation of Newton's *Opticks* and by the nineteenth century the hard core of distrust of speculation was prominent as a positivist movement.

This book therefore brings together a study of the development of science in two dimensions—in place and in time. One could add further dimensions—distinguishing for example disciplines within science, but our aim has been to provide a general overview supplemented by various insights into the emergence of science in different countries. When the time comes to write a full comparative study we hope that our endeavours here will be of use to our successors. Meanwhile each of the contributors has provided a valuable characterisation of the science which emerged from a particular cultural situation.

The chapters of this book are based on papers presented at the summer meeting of the British Society for the History of Science at Leeds in 1974. Plans for publication have been complicated by the fact that immediately after organising the conference in Leeds, I moved to the University of Kent at Canterbury. I would like to thank contributors for their co-operation particularly in the planning stage of this book, when ultimate publication was uncertain. In addition to the speakers at the conference, I should like to acknowledge the services of the chairmen: Professor D. S. L. Cardwell, Dr J. A. Chaldecott, Dr E. G. Forbes, Professor A. R. Hall, Dr J. R. Ravetz and Mr G. L'E Turner. If space permitted I would also list more than a hundred names of those who attended the conference and took part in questions and discussion (concluding in a general discussion with introductory remarks by Mr J. B. Morell). My sincere thanks are due to all those and especially to my former colleagues from whose constructive comments I have been able to benefit. Mr R. G. A. Dolby and Mr M. Jack of the University of Kent kindly read through this introduction and made valuable suggestions for improvement. I should also like to thank Mrs Veronica Ansley for her help with the index. Finally my thanks are due to the Nuffield Foundation for generous support of my project to study aspects of the development of science in a Western European context, of which this survey forms a part.

Canterbury, November 1974.

References

1. In other terms the foundations of modern science were laid not only by Newton and Boyle but also by Galileo and Descartes, Lavoisier and Liebig, among many others.
2. J. G. Crowther (1967), *The Social Relations of Science*, London, pp. 227–8. For another example of the argument of economic determinism see E. Zilsel (1957), "The Origins of Gilbert's Scientific Method". In *Roots of Scientific Thought* (eds. P. P. Wiener and A. Noland), New York, pp. 219–50.
3. Preface to *The Chemist*, **i** (1824).
4. "On the electricity excited by the mere contact of conducting substance of different kinds", *Phil. Trans.* (1800), 403–31.
5. Quoted by Sir Harold Hartley, *Humphry Davy* (1966), p. 86.
6. J. von Liebig (1851), *Letters on Chemistry*, 3rd edn., London, p. 26.
7. *Ibid*, p. 96.
8. The use of Latin by Linnaeus in botanical nomenclature has persisted in the twentieth century.
9. S. Shapin, "The Pottery Philosophical Society, 1819–1935: An examination of the cultural uses of provincial science", *Science studies*. **ii** (1972) 311–36. D. S. L. Cardwell "Science in Manchester in the Nineteenth Century" (Broadcast talk, 1973). A. Thackray "Natural knowledge in a Cultural Context: The Manchester Model" *American Historical Review* **lxxix** (1974), 672–709.
10. *Reflections on the Decline of Science in England* (1830), London.
11. H. T. Pledge (1939), *Science since 1500*, London, p. 10. Pledge intended to undertake a full study of the problem but this was interrupted by World War II and twenty years later he admitted that he had been unable to follow up his earlier investigations (*ibid*, 2nd edn., 1966, Preparatory note, p. 3).
12. On "Irish science" and "Welsh science" see the comments by D. S. L. Cardwell (1972), *The Organisation of Science in England*, 2nd edn., London, pp. 7–8.

13. D. S. L. Cardwell (1972), *Technology, Science and History*, London, pp. 9, 27.

14. In the seventeenth century the major work of drainage of the fens of the Nene and the Ouse was directed by the Dutch engineer, Cornelius Vermuyden.

15. Henry Briggs (c. 1560–1631), professor of geometry, Edmund Gunter (1581–1620) and Henry Gellibrand (1597–1637), successive professors of astronomy at Gresham College, were all very much concerned with problems of navigation.

16. R. Porter (1973), "The Industrial Revolution and the Rise of the Science of Geology". In *Changing Perspectives in the History of Science* (eds. M. Teich and R. Young), London, pp. 320–43 (333–4).

17. *Histoire des Sciences et des Savants depuis deux Siècles* (1885), 2nd edn., Geneva and Basle; see especially pp. 372–487.

18. R. K. Merton, "Science, Technology and Society in Seventeeth Century England", *Osiris*, **iv** (1938), 360–632. Reprinted (1970), New York.

19. See the chapter by P. M. Rattansi. See also A. R. Hall, "Merton Revisited", *History of Science*, **ii** (1963), 1–16.

20. *The Personal Papers of Anton Chekhov* (1948), New York, p. 29.

21. When the Earl of Bridgewater died in 1829 he left a sum of £8000 to be paid to persons selected by the President of the Royal Society to write a number of treatises "On the Power, Wisdom and Goodness of God, as Manifested in the Creation". Nine treatises were published in the years 1833–37 and were popular enough to go through numerous editions.

22. That the anterior roots of the spinal nerve are motor and the posterior sensory.

23. Sir Gordon Gordon-Taylor (1958), *Sir Charles Bell, His Life and Times*, Edinburgh and London, p. 128.

24. *Ibid*, p. 129.

25. T. S. Kuhn (1962), "Energy Conservation as an Example of Simultaneous Discovery" (ed. M. Clagett), *Critical Problems in the History of Science*, Madison, pp. 321–56. Y. Elkana (1974), *The Discovery of Conservation of Energy*, Cambridge, Mass.

26. "English Physics and the Mechanical Model". In (1914, 1962) *The Aim and Structure of Physical Theory*, New York, Atheneum, pp. 69–80. By "English" Duhem means "British".

27. David Knight has kindly reminded me that Young also studied in London, Edinburgh and Cambridge. This experience makes Young a valuable witness in a comparative assessment of university education in England, Scotland and Hanover. See for example A. Wood (1954), *Thomas Young, Natural Philosopher, 1773–1829*, Cambridge.

28. I. Brunel (1971), *The Life of Isambard Kingdom Brunel, Civil Engineer (1870)*, Reprint, Newton Abbot, p. 5.

29. *Elements of Chemistry* (1790) (trans. R. Kerr), Edinburgh. Dover Reprint, New York, 1965, Preface p. xxiv.

30. Some of these ideas were analysed by K. J. Boughey (1972), in an unpublished Ph.D. thesis: *Studies in the Role of Positivism in Nineteenth-Century French Chemistry*, University of Leeds.

31. *Journal für praktische Chemie*, **xv** (1877), 473. G. Bruni (1913), "The work of J. H. van't Hoff", *Annual Report of the Smithsonian Institution*, pp. 771–2.

2. Mathematicians, Mechanics and Experimental Machines in Northern Italy in the Sixteenth Century

A. G. KELLER
(University of Leicester)

I   The Italian Scene

Italy in the late sixteenth century looks to modern eyes and especially to those brought up in the modes of thinking of contemporary science, a very strange and alien world. The beliefs and practices of magic were still very much alive; society was hierarchical and its secular ideals were those of honour and display. Personal violence was common, and could not be suppressed even by the most savage and cruel punishments. Even if these observations could certainly be held to apply as much, or more, to most other parts of Europe, the love of display, and of the bizarre and the ornate, seems to have been even stronger in Italy than elsewhere. Likewise the Italian taste for dramatised conflict and private combat made late Cinquecento Italy appear to be a continuous costume melodrama to her neighbours—that is why they so often set their gorier plays in Italy.

Yet the Italian landscape presents other features, less Gothic and more prophetic. Although at least one Northern visitor held up his hands at the wretched quantities of meat consumed per annum in an average Tuscan city, his own account of the diversity of vegetables in the Italian "salad" (then a novel Italian idea) suggests that the diet of the middle and lower classes in Tuscany and the Po basin was probably rather better than elsewhere.[1] The Italian climate ensures that crops seldom fail through excess of summer rain. New cereal crops, such as rice introduced from the Middle East in the fifteenth century and maize introduced from the Americas in the sixteenth century, provided a motive for bringing fresh lands, unsuitable to wheat, under

15

cultivation and so helped protect the mass of the population from famine. The very smallness of states like Tuscany and Venice made it easier for them to try to safeguard the public health from plague (by quarantine, health certificates) and to maintain a constant food supply (by official importation in years of shortage). Despite the elaborate structure of feudal forms in language and symbol the social pattern already showed some signs of progress. The abyss that gaped as wide between rich and poor there as elsewhere could not conceal the relative weakness of an ancient hereditary nobility. The Grand Dukes of Tuscany were the great-grandchildren of bankers; the ruling class of the Venetian republic were merchants and proud of it. It was somewhat easier for exceptional talent to rise in the world, and so to win not only wealth and power (which could be done elsewhere) but respect—which is harder to attain. The "artist" had succeeded in distinguishing himself from the artisan. Poor boys who were good at painting or carving or at any kind of metalwork or ceramic, at the making of ornate furnishings and utensils for the palaces of the rich—and also instruments and machines, provided they were ingenious enough—could hope that their talent would be discovered and rewarded.[2]

However, such men were naturally few. More numerous than those who wished to rise were those who were keen to stay where they were—near the top. The old feudal landed nobility had always seen their social function as fighting and this was now not only the path to honour, but also to the maintenance of their social position (and its income). Those of mathematical or intellectual bent were led into the rapidly growing branches of artillery and military engineering. The introduction of cannon led in the last years of the fifteenth century and the early part of the sixteenth to a revolution in the art of war, which made it necessary to recast the whole system of fortification. Instead of castles scattered across the country to hold a large number of strong points, towns were the chief units of defence; a few commanding positions ordered the grand strategy of a state's defence. But mere walls would serve for little against cannon. A new system evolved, to combine maximum field of fire from ramparts that were primarily gun platforms, with minimum vulnerability.[3] The design of these works required the services of a large number of men possessed of mathematical skill. Artillery and fortification thus provided a new career open to talent; if there was no employment at home at any given time, in fortifying towns and castles, it might be found across the Alps. "If anyone is an engineer (*mechanicus*) or an architect building palaces for kings or a painter . . . *Italus est*" wrote the French scholar Ramus in 1567, advocating that his fellow countrymen study mathematics so as to catch up on Germans and Italians. The pursuit of mathematics would lead to technological, and therefore to military and economic supremacy. At the Venetian Arsenal, for instance, all the skilled men had a good grounding in mathematics, and to this Ramus ascribed its efficiency.[4] Because the skills required by the new military technology could be found in all classes, it opened the way to a titled but genuine meritocracy. Such was the story of Captain Francesco Marchi, who started life as a courier, and claimed he did not learn to read until he was thirty-two.[5]

At the same time, the urban patricians were moving into agriculture. When commerce flagged, the wealthy merchant invested in the security of land, which brought with it social status. But this land was not an inheritance handed down from remote ancestors, and he might well bring to his estates something of the values of the trader; he could hope to increase its profitability by agricultural improvement, by the introduction of new crops, by the drainage of marshland, or by irrigation to turn arable into water-meadows. Through the later Middle Ages and the Renaissance a network of channels and leats was dug and embankments raised to form the very distinctive landscape of the lower Po basin as it is today. The principal towns of the basin lie on the roughly parallel tributaries of the Po; they too were linked by navigation canals. Thus a school of hydraulic engineers was trained, often interchangeable with the military engineers.

II   Theoretical Arts in the States of Northern Italy

It hardly needs to be said that no such thing as "science" had then emerged. Nor indeed did Italy then provide a "national context", divided as it was into several small states with differing constitutions, economies and traditions. Pope Julius II proclaimed his battle cry: "Barbarians Out". Machiavelli echoed his words, and so did many others. But in the second half of the sixteenth century the Habsburg barbarians were as strongly entrenched as ever in many parts of the peninsula, and Habsburg power overshadowed the rest. Just as *Italy* meant a cultural and geographical concept, so we could look at all the cultural activities that were to give rise to science. Several features in Italian life could be regarded as the seeds of a future flowering of science; the first snowdrops had shown their heads above surface, but their significance lay rather in what they suggested about what was going on underground, than in their intrinsic effect.

Yet if we were to examine everything that could be so treated—where would it end? Knowledge was being accumulated and techniques introduced in almost every walk of life. We could speak of the work, the methods and ideas of architects, painters, dyers, potters, glaziers, workers in iron and brass and bronze, just as much as improving landlords who experimented with seed-drills, or apothecaries with their Asian and American drugs. Instead of trying to survey the whole huge field, I propose to choose a "growth point" that seems important for the immediate future. Two types of activity were pre-eminently seen to be compounded from theoretical knowledge (*scientia*) and practical operation (*ars*): the various applied mathematical arts, and the study and use of "simples"—as it were, applied natural history. Since both types of activity had so many and such complex ramifications, I propose to be more specific still, and deal with one particular mathematical art, where the confrontation of theory with experiment was to prove the pivot on which seventeenth-century physics turned; a group of letters illustrates a crucial moment in the story, for they related both the motives for a translation of an important original work in mechanics into Italian, and also the attempt to check its theories experimentally.

The new mechanical sciences of the early seventeenth century were to prove the "cutting edge of objectivity", the door step that led into the new world. The study of natural history takes a more humble position in our estimation. Outside anatomy there were as yet few startling discoveries to be made. What later became biology and geology were still at the fact-collecting stage, which can easily be ridiculed as dull, pedantic, fit only for lesser unheroic minds. Nevertheless, changes were under way in these fields too. Such historical research as we have had so far has concentrated on the medical faculties of the universities. But another kind of institution, later very important, effect-ively begins in the late sixteenth century—the natural history museum. Cabinets of exotic curios long preceded these organised collections, of course, but the voyages of discovery in the Renaissance excited interest in the outlandish but profitable mineral and plant products of distant lands unknown, or little known, to the Ancients. It would be a valuable exercise to essay a history of such museums as those of Imperato and Calzolari; here too there is a fruitful opposition of theory (inherited mainly from Dioscorides and Galen) with practice. There was also an opposition between utilitarian and aesthetic motives. Both of these men were apothecaries by trade, con-cerned with the medicinal uses of their specimens and both delighted in the marvellous and in the stimulus of strange sights and odours. In the task of identification and systematisation, of sorting the genuine from the spurious, and establishing relationships, they certainly fulfilled a major historical function.

III   Tartaglia and the Interest in Mechanics

Whatever the explanation, it would seem certain that there was greater interest in mathematics, and still more in the applications of mathematics in the mid-sixteenth century than in previous generations.[6] These years saw the appearance of a new profession: the teacher of mathematics who might also survey lands, design fortifications, invent new mathematical instruments and draw up maps and plans. He usually taught perspective and perhaps the mathematical side of architecture, as well as cartography and, in maritime countries, navigation. His instruments were intended to simplify geometrical or trigonometrical rather than arithmetical calculation. There had been practical mathematicians and mathematical instruments before: Italy was by no means in the van in this respect. Although in the early years of the mathematical instrument industry, such as in horology, Italy did produce several pioneers (for such families as the Volpaias and the Dantis proved the equals of any in Europe), the majority of instruments were imported. South Germany led all Europe. Nevertheless, the general mathematical teacher/ practitioner, who primarily taught the geometrical arts and could be consulted on problems connected with them, and for whom astrology was at most a sideline is at first an Italian phenomenon, as his clientele is largely due to those features of Italian life that have been outlined above. Like Italian military engineers, painters and architects, Italian mathematical practitioners travelled across Europe in search of employment. Indeed, the teaching of mathematics might be a suitable peace-time activity for a military engineer. The traffic was not, of course, in one direction only. Drainage problems were a Dutch speciality and instruments and locks continued to come from Germany. But one man imposed his view of the mathematician, his task, his capabilities, his social function, upon Europe, and that man was an Italian: Niccoló Tartaglia.[7] Tartaglia was another self-made man. In a passage inserted in his *Quesiti*, he tells how his father had been a poor courier, who left Niccoló an orphan at the age of six, and his mother "liquida di beni della fortuna". While still a child he received a dreadful gash on the face at the siege of Brescia, but there was no money to buy ointments, or go to a doctor (probably that did Niccoló good, for his mother contented herself with keeping the wound clean).

> "Before my father died, I was sent for some months to a reading school . . . at that time I was very small, that is at the age of five or six . . . later being fourteen years old or thereabouts, I went willingly to a writing school . . . in which time I learnt to do my abc as far as letter k of the merchant hand." "And why as far as letter k and no further? . . ." "Because the terms of payment were to give one third at the beginning and another third at letter k . . . and at the said point there was no money to pay what was due."[8]

So Tartaglia had to teach himself even the second half of the alphabet. He taught foreign visitors, such as Wentworth from England and Hurtado de Mendoza from Spain; his books were translated into German, English and French, while the influence of his nearest foreign counterpart, Apianus, was confined to the German-speaking lands. Tartaglia's name was cited more than that of any other writer on applied mathematics, not to mention borrowings that were not acknowledged. He introduced fresh arts into the mathematical family, and with great boldness and verve in 1537 proclaimed to the world a New Science—ballistics.[9] The frontispiece of his book has been analysed more than once, for although the theme is obvious there are one or two puzzles if we try to read the symbolism as intended in every detail.[10] Among a host of mathematical arts, led by Tartaglia in person, which greet the new student in the forecourt of the castle of knowledge, the two foundation studies Arithmetic and Geometry flank the teacher; in the next rank stand the other two quadrivial arts, Music and Astronomy, now joined by Perspective and Architecture; behind them stand a crowd of "natural magic" arts. In the introductory dedication to his translation of Euclid (1543), which reappears slightly modified in later editions as two public lectures on mathematics, Tartaglia discusses the question of the number of mathematical arts, and names those studies which he holds to be based on mathematics.[11] The natural magical ones, which it turns out he took from Cornelius Agrippa, are now summarily brushed aside in a sentence. Among the more realistic mathematical arts, there is one we have not yet mentioned.

In the seventh book of his *Quesiti et Inventioni Diverse* (1546), Tartaglia presents us with the following dialogue, between himself and the Imperial Ambassador to Venice, Don Diego Hurtado de Mendoza, who had arrived there in 1539, so the conversation may be imagined to have taken place in the early 1540s.

> THE LORD AMBASSADOR: Tartaglia, since we took a vacation from our readings in Euclid, I have discovered some new things on Mathematics.
> NICCOLÓ: What has Your Lordship discovered?
> L.A.: The Mechanical Questions of Aristotle, Greek and Latin.
> N.: It is quite a time since I saw them, particularly in Latin.
> L.A.: What do you think of them?
> N.: It is very good, and certainly these things are very ingenious, and of profound learning.

The ambassador agrees, although he explains that he has some difficulties which he would like Tartaglia to explain, to which the latter replies that there are indeed weaknesses in the method and procedure of the Mechanical Questions, which can be improved by means of the "science of weights", which he then proceeds to expound (without acknowledging his debt to Jordanus).[12]

Presumably it was Don Diego who influenced Tartaglia to take the *Mechanical Questions*, or *Mechanical Problems* seriously. The Italian Euclid of 1543 is dedicated to Gabriele Tadino, Prior of Barletta, and later commander of the Spanish artillery, who is reminded by Tartaglia of all the geographical arts that depend on mathematics, which the two of them had often discussed "upon your navigation chart and your German Globe". But he must also mention mechanics "the cause of every ingenious mechanical invention", based on the principles of Aristotle's *Mechanical Questions*, which is in turn derived from the science of weights—now ascribed to Jordanus.

Don Diego took the matter further.[13] About the same time he persuaded another scholar, Alessandro Piccolomini, then at Padua, to produce a paraphrase, or rather exposition of the *Mechanical Questions*. He refers to this paraphrase in a work that appeared in 1542 (although the paraphrase was not in fact published until 1547), and he also composed his own translation from Greek into Spanish, in 1546. Although Piccolomini frequently links his study of the theory to the actual mechanical practice of the time, and supposedly thought it would be useful to have an Italian version for architects and engineers without Latin, the translation that was made did not appear until 1582.

Indeed, apart from Tartaglia's *Travagliata Inventione*, where he proposes a method of re-floating wrecks which he derived from Archimedes' hydrostatics, and his translation of Jordanus' *De Ponderositate* which his printer, Curtius Troianus, brought out together with new editions of his Euclid and the Piccolomini paraphrase in 1565, nothing seems to have been published in the way of mathematical mechanics in Italy for over thirty years. It could be that those men with talents combining mathematical ability with a utilitarian approach, who might therefore have been interested in such problems, were diverted by the immense and rewarding task of fortification which was now at its height. The art of war was always acclaimed the noblest of arts. Perhaps it was necessary for mechanics to be shown to possess a military purpose in order that it be considered important.

IV   The Mechanics of Guidobaldo dal Monte

The idea of a new mechanics was revived by Guidobaldo de'Marchesi dal Monte, who was a man like Hurtado de Mendoza—an aristocrat with a military background.[14] In younger days he had fought against the Turks, and was later inspector general of the fortifications of the Grand Duchy of Tuscany; in 1580 he was assured that "your valour in Arms and Chivalry are celebrated by all". He was of a more retiring character than Hurtado de Mendoza— the scholar. He had attached himself to the school of learned humanist-mathematicians which gathered round Federico Commandino at the court of the Duchy of Urbino. Guidobaldo's treatise on mechanics is among those recently translated, with an excellent summary of his views, by Stillman Drake.[15] Although it is hard to accept Professor Drake's notion of a purist "Central Italian" school, as opposed to the practical men of the north of Italy, it is true that Guidobaldo was a purist in his concern to establish mechanics as a branch of rigorous axiomatic geometry. All the same, in the dedication of his *Mechanica*, he stresses the utility of his subject as the source of all techniques, which arise like a great river from a little spring which has by now swelled to a mighty stream of different trades, all employing machinery: in agriculture the plough, in commerce waggons and ships; swipes, presses, cranes, saws, weapons of war; and all the tools of "those who work with wood, stone and marble, wines, oils and unguents, iron, gold, and other metals, as well as surgeons, barbers, bakers, tailors" and many more.

The *Mechanica* was Guidobaldo's first book, printed two years after his master Commandino's death. It may be that his choice of title is one reason why we still use this name for the science of mechanics, which has long dealt with a far wider range of subjects than the comprehension of how machines work—instead of some name taken from the Greek or Latin word for motion. In his choice he may have been influenced by the title of the *Mechanical Questions* (often known simply as *Mechanica*) as well as by Heron's treatise of the same name. What Guidobaldo was trying to do was to answer the problems set by pseudo-Aristotle, but give them the cogency that could only be supplied by rigorous proofs based on Archimedes' *On Equilibrium of Planes*, proceeding from the balance to the lever and thence to the rest of the six Simple Machines explained by Heron, as later excerpted and summarised by Pappus in his *Mathematical Collections*. In his Classical spirit, Guidobaldo rejected the advances that had been made by the Jordanist school, as well as the work of Tartaglia. Now Archimedes was clearly understood, and could be supplemented by Pappus; it therefore must be possible to by-pass the older writers on these topics altogether. Guidobaldo was evidently confident that his mathematical reasoning would be confirmed by experimental test. Even if he himself was probably not very interested in the improvement of machinery, he was satisfied that what was proved geometrically would work in the real world.

V   The Machines and the Tests of Savorgnan

At this stage, the initiative in this conscious effort to apply theory and experiment to machinery was taken up by a third figure. Count Giulio Savorgnan was also of the landed nobility, lord of Osoppo, count of Belgrado, and had other possessions on the north-east frontier of the Venetian Republic, which his father and grandfather had defended against the might of the Habsburgs.[16] At the age of sixteen he had been sent off to war, but whereas men like Hurtado de Mendoza and Dal Monte engaged in campaigning (as did Descartes and so many others) as a form of higher education, he spent most of his life as a soldier. For some thirty years he was in the service of the Republic in its eastern outposts, at Zara and Lissa on the Dalmatian coast (he was governor of Dalmatia in 1569), on Corfu, in Crete and in Cyprus. He was first and foremost a military engineer, responsible for new fortifications in almost every town at which he was stationed. He is credited with the construction of some fifty bastions, which seems not unlikely. Like Don Diego he was of liberal temper, with sympathies for all classes of men and an independent mind. In the mid-1570s, after the battle of Lepanto, he returned to Osoppo and busied himself with re-fortifying his castle, and constructing a new aqueduct from the local river Tagliamento, to provide flood relief.

He was of the same generation as Mendoza, and had been Tartaglia's student too. He appears in the *Quesiti*, asking a simple question about how to gauge the weight of cannon balls.[17] But Tartaglia dedicated to him a free translation and commentary on Archimedes' *On Floating Bodies* Bk. I, published as the second part of the *Travagliata Inventione*. There he explains that Savorgnan had some time previously given him a list of no less than twenty-nine queries, which displayed his "ingenious curiosity at seeking out the secret effects of nature" (curioso ingegno nel ricercare li secreti effetti di natura) ; only two are reported, on specific gravities and hydrostatic balance.[18]

Old as he was, Savorgnan was not left to rust in his retirement. Often called away for consultations, he was brought back into office in 1587 as "superintendent-general of artillery and fortifications", and in the 1590s when already octogenarian he presided over the planning of that mathematical model incarnate, the new town of Palma Nova, which the Most Serene Republic of Venice laid out in the last years of the sixteenth century, to guard her serenity on the eastward side. Not that he had been idle in the interim, for he turned his house into a centre of activities which combined the traditional military leisure pursuits, riding, hunting, the old knightly life, the "vita cavalleresca", with intellectual activity ; his house became a "resort of talented people, a hostel for soldiers and scholars" (ridotto di persone virtuose ed un albergo di soldati e dottori). There he became acquainted with Dal Monte's new book on mechanics, and decided to have it translated into Italian.[19] The man selected for the task was Filippo Pigafetta, a professional author, not specifically interested as yet in such problems, although he was later to write one of many pamphlets on the shifting of the Vatican obelisk. Because this was a sponsored translation, Pigafetta's dedication to Savorgnan holds much greater interest than the common run of obsequious dedications as it portrays Savorgnan as he wished to be seen and enlarges on his motives. Much of this dedication has been translated in Professor Drake's *Mechanics in sixteenth century Italy*, but with a number of omissions (not always indicated) of just those passages which provide the social context.[20] He does include a key passage on Savorgnan's armoury, "in the fashion of a store of arms all kept well polished in their places" (a guisa d'una bottega d'arme pulitamente a suoi luoghi servate), which also contained a

> "magazine of . . . machines to move wights, of which you have through your industry constructed perhaps a dozen different sorts, some to draw weights along, some to raise great weights with little force. One of them has but a single toothed wheel, yet it pulls up five of your cannon vertically by the strength of Gradasso your dwarf. Another with but an ounce of force placed on the handle sets in motion fourteen thousand pounds in weight; and since a man usually has fifty pounds of force in his arm, if this were used on the same handle, it is evident that the said machine would have the incredible power of moving more than eight million pounds."

Pigafetta then goes on to explain the point of all this machinery : Savorgnan's engines can be transported on the back of a mule, "some even by a man".

They are particularly useful for "handling and transporting great pieces of artillery". And here we come to the nub: a new use for machines, which made them of interest to these military gentlemen with their new-found training in geometry.

The mobility (relative to earlier times) of sixteenth-century warfare, the demand for rapid deployment of heavy artillery, required the use of lifting gear that would not only be powerful, but also light. Screw-jacks and ratchet-jacks were probably invented in the Renaissance for a purpose quite similar to their present one, either to support one side of a vehicle while a wheel is fitted, or to right it if it tips over.[21] But they were diffused in the late sixteenth century largely as a piece of military equipment, to raise the tail of a cannon so as to adjust the angle of elevation. Larger instruments were developed from the hoists that had long been used by builders to lift heavy stone blocks. But as they had to be easily portable, they had to be much smaller and lighter, and yet not less powerful, if they were required to hoist a gun on to its carriage, or rescue it from a muddy ditch. Pigafetta tells two stories in explanation. One occurred in Savorgnan's first war, in 1529, when a French unit was withdrawing from Milan: a heavy cannon fell, broke through a culvert and lay in the mud; and "not having an engine to pull it out, so much time was consumed" that the enemy caught up with them and captured the whole force. More recently another French army in Italy was manoeuvring into position, when a cannon slipped from its carriage; the whole body of men stopped, the officers dismounted, and again much time was lost before they could get it back on its wheels. Luckily for them this time their enemy did not take advantage of the delay.

Of course, equipment to meet this kind of contingency could only be ancillary, but it would still be of great importance to have the best available. Other writers on the art of war at the time display their own inventions to the same effect, and Figure 1 shows the device of a contemporary French engineer, Jean Errard. Two north Italian engineers, Bosca (whose *Dell' Espugnatione et Difesa delle Fortezze* was published in 1585, the year after Errard's book came out), and Gentilini (who published *Instruttioni di Artiglieri* in 1598), both discuss the problem but without illustrating their solutions.[22] The difficulty involved in hauling cannon up a steep incline or over a rocky track was somewhat different: this too attracted the attention of several authors. A third type of larger engine could be used in the armoury; these of course did not need to be portable.

This was Savorgnan's motive for taking such an interest in mechanics, and for trying to develop his own devices. Pigafetta's dedication would in itself be a valuable indication of the reasons behind this renewed interest in the subject. But there is more. A group of letters is preserved in the Biblioteca Ambrosiana, Milan.[23] Three of these are from Savorgnan to an unknown person, possibly the amateur of all the arts and sciences and correspondent

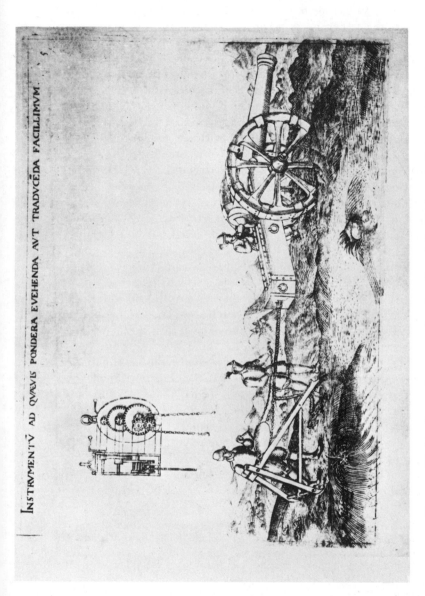

Figure 1   Machine for rescuing cannon; from Jean Errard, Premier Livre des Instruments mathématiques mécaniques (Nancy, 1584)

of Galileo, Giovan Vicenzo Pinelli, from whose collection they passed to the
Ambrosiana. More than half the second letter and the first section of the third
deal with the experiments which lie behind the machines described by
Pigafetta. The fourth letter in the file is from Pigafetta to Guidobaldo,
presenting a number of queries about the translation; there are four sheets
of corrections and comments. There are two letters from Guidobaldo; the first
is his answer to this letter of his translator, the second in answer to another,
which has not been preserved here. Savorgnan mentions Guidobaldo's theory
of the pulley in his second letter; it is evident from references in the interchange
between Pigafetta and Guidobaldo that Count Giulio tested Guidobaldo's
theories, and sought to apply them in his inventions.

   The remainder of Savorgnan's three letters discuss the international situation, fraught with peril as it always is. Given his personal history and his
geographical situation Savorgnan was very concerned about the Turkish
peril, and Italian unreadiness to meet it. His first letter, of 29 May 1578, is
devoted to this entirely.[24] But the second, dated three weeks later, 17 June,
describes the arrival of "your perpetual screw", which Savorgnan has fitted
up, and compared with his own "Archimedean instruments", equipped with
"toothed bar on one side only" (rack-bars?), which like his correspondent's
can raise forty-five thousand pounds "if rope or teeth can withstand it".[25] For
his friend's device, he has made a wooden bar toothed on both sides, and
"you will see a fine effect, for when the screw is turned, the wooden bar rides
up quicker than mine do". However, he hastens to point out that this
advantage is out-weighed by its lack of strength, "just as honour and avarice
cannot go together, so it is not possible that in one instrument there should
be great strength, and velocity". Indeed, he has found rack-bars inconvenient
in general because of the limits on their length, and so on the height of their
lift, not normally above 4 to 5 feet. Savorgnan's "machine of iron teeth in a
round wheel" is very effective as a cannon hoist, limited only by the length
and strength of the cable. Unfortunately it is rather large, "five foot long by
one and a half feet high and wide"—which does not sound very big, unless
it is recalled that these machines had to be easily portable. To solve this
problem, he designed another:

> "just as of a mare and an ass mules are born, composed of two natures, I have
> had made a machine with wheels in the manner of Archimedes, and at that point
> where the long toothed iron bar is applied, I apply a cord wound round the
> drum . . . this machine mixed of two natures has this other happy feature, that
> it is only one foot two inches long, and only seven inches wide, and is turned by
> hand, so that a force of one pound raises me 300; 50 would raise fifteen thousand,
> with one hand alone, and so with such a little instrument, provided the rope
> supports the weight, the force is so great that a rope like the one normally used
> to do justice upon men, applying them as the weight, becomes a little thinner
> than a finger of a man's hand".

And thus Savorgnan reached his objective: "I could put a greater weight on it than on the tackle of signor Guidobaldo which passes over four or six wheels, with which one could almost pull the mountain of Osoppo as far as Belgrado".

On the other hand, the perpetual screw—so popular with inventors of the time—does not seem serviceable because of its lack of force; "it is a really fantastic oddity"—good only for learned arguments. "Let Your Honour test it with some worthy men of understanding and try to learn from them what use it could be, I would take pleasure in knowing it—meantime, perhaps if the French invade again, we could use it to haul their whole army out of the country. . . ."

He then returns to an assessment of the international situation. As usual he is pessimistic.

> "Preachers are always shouting about the evil that will come upon us for our sins, there is never one to be found who when he preaches prophesies happiness; I seem to have become one of these . . . and I despair of finding a way to fortify Osoppo better and faster . . . so as to be the last of all, as my ancestors were the first to serve this Illustrious Dominion."

The third letter was written almost two full years later, on 8 June 1580, and Savorgnan was feeling more cheerful.[26] Most of the letter is taken up with a discussion of his correspondent's report of conflict between Turkey and Persia, which naturally pleased them, but he begins by thanking him for another gift of

> "metal rollers . . . very dear to me because I have learnt from them the fashion of having others made that way, with very little metal, trim and strong. . . . On the subject of rollers, I would let you know, that with a force of ten or eleven ounces I draw 22 000 pounds, in rivalry with great Archimedes. I believe that even when tired I would pull by hand, with 50 pounds, in such a way that more than 600 000 pounds weight would be pulled, which if Archimedes knew, he would despair and I would laugh at his despair, and if Your Honour goes to Sicily, as I have understood you have some thought of doing, please give my regards to the walls of that city in which that worthy man stood,"

although, he adds, with the plague raging there, "you will be at risk of going to talk to Archimedes himself, and not just his walls".

He was much happier too with the state of Osoppo, and in view of Pigafetta's description the next year, it is interesting to see him inviting his friend over to "a house furnished with everything, and liberty to study, to walk and to live as you like, without having to pay court to anyone". Recording the success of "my small pieces of artillery", he remarks that "they will also please Your Honour in respect of some trestles (or trolleys?—*cavaletti*) with rollers underneath in odd fashion, fine and convenient, which I have had made".

Following these letters in the file is that from Pigafetta to Guidobaldo, sent from Padua, 5 November 1580, just five months after the last of Savorgnan. He may have been staying with Pinelli for he refers to "a certain ancient

Tuscan writer" (i.e. on geometry) "seen by me in the library of Sr. Pinello". He had evidently written before, and Guidobaldo had replied, enclosing "Advice about translating" (preserved with the letters), in which he argued in favour of Italianising Latin technical terms rather than attempting to give new meanings to old Italian words.[27] Guidobaldo recommends the works of Egnazio Danti, and his old master Commandino's translation of Euclid as models of how to proceed. Much of Pigafetta's answer is concerned with this linguistic discussion, very interesting in its own right.[28] Some of the points reappeared in his commentary when the translation was published. But he also enclosed three sections of the book newly completed, and asked some questions, particularly about the theory of the operation of pulleys, which he had not understood. In the first and longest of these, he explains how

> "while I was at Osoppo with the Illustrious Sr. Giulio Savorgnan, who was the cause of this translation, every day with that gentleman, who does nothing else but handle military constructions, tests were made on Pulleys, of which he has plenty, of different kinds, very finely wrought by a German of his called Master Fait. And we found that in fact they did not work in the way that reason shows. Because, for example, he has two sheaves with three pulleys each, when the rope is wound round them all and two weights attached, one, that is the heavier served as the weight, placed in the lower sheave, and the other, lighter, for the power; the pulleys and the rope well greased first with soap. Then according to the demonstration set down on page 74b, the power comes out at one-seventh of the weight, that is if the weight A is 700 pounds, the power of G will sustain it with 100. But as I have said, in fact the power did not answer to the weight, nor experience to reason . . . we believed this happened through the resistance of the pulleys, axles and ropes, and through their heaviness (*gravezza*); of which nevertheless that gentleman desired of you to know the cause, if there be any other."

Guidobaldo wrote back promptly from Pesaro, on 14 November;[29] he observed that he had just written at length to Giacomo Contarini "on the way the experiments ought to be made", as he had raised very similar doubts about Guidobaldo's theorems. But Guidobaldo is dubious whether Savorgnan's experiment can have been carried out as stated, for "l'esperienza per a punto si confronta con la ragione"—he explains.

I mean that it is simply not possible to reproduce his illustration on page 74b as an experiment: it is only an ideal situation: and he suggests an alternative system which is shown in Figures 2 and 3. And then the experiment will work perfectly, for he claims that the pulleys and their axles only offer resistance to upward motion, but not to a state of equilibrium "as I have proved many times".

Although in this letter Guidobaldo requests that there be no suggestion on the title-page that he had overseen the translation, it is quite clear that he did so, for each part was submitted to him, and we have some six pages of corrections and definitions on lines similar to those already examined. Pigafetta must have replied, but all we have in this file is a further answer from

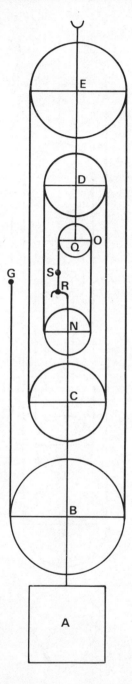

Figure 2   Woodcut from printed version of Guidobaldo's *Mechanica* (1577)

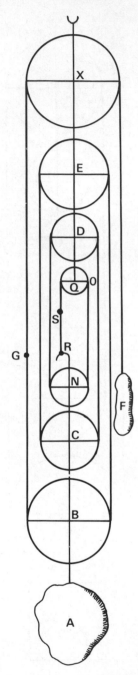

Figure 3　Guidobaldo's drawing, sent to Pigafetta (1580)

Guidobaldo, dated 21 December. Pigafetta had enquired about a piece Guidobaldo had formerly written on the fortifications of Corfu. Perhaps he thought there would be a better market for this more popular subject. Guidobaldo brushed him aside—it was but "una bagatella", nothing serious—although he still knew where it was to be found—"and if Sr. Giulio Savorgnan should see it, please make my excuses". He added that he was anxious to know if Savorgnan was still unconvinced on the pulley problem, as Contarini was, "who wrote to me a few days ago, that the axles of the sheaves make resistance both in sustaining and in moving, and that experiment in this regard is not conformable to the demonstrations that are in my book". And Contarini at any rate believed that Savorgnan was of the same view, since "many times they have made the experiment". So Guidobaldo asks his translator to make them comprehend the point, "as Your Honour has undoubtedly understood it very well; and as I am not able to do anything else, I am resolved to make some models and send them", to clarify this point, "which in machines is quite fundamental". Although models were then regularly used to demonstrate the virtues of particular machines, this seems to be the first case where it is proposed to build a model specifically for experimental purposes. For his part, Savorgnan had also been busy, as Pigafetta mentioned that "Sr. Giulio has discovered empirically another way to recognise the force of the screw". Guidobaldo expressed his pleasure, but insisted he would like to see it demonstrated and know how it worked. In that case perhaps it could be included in the book as an appendix to the section on the screw—but not otherwise. Indeed, "once I saw the experiment, perhaps I could from that alone go on to discover the proof".

It would not be easy to reconstruct from this sparse information how the device to measure the force of a screw was to function, any more than one could work out the exact nature of the machines described in Savorgnan's own letters, without a drawing. Spur-gears, worm-gears, rack-and-pinion, block-and-tackle, winch and pulley appear in different combinations in all the ingenious lifting apparatus of the sixteenth century. Whether Count Giulio's Mule was in reality so very revolutionary we cannot now know. It is not likely. The experimental objections of Savorgnan and Contarini to Guidobaldo's abstract statics are at least on the right lines, even if they could only infer what was wrong, but not how to make it right. But if we would understand the nature of the ground from which great new ideas in mechanics were to sprout in the next generation, I suggest we would do well to look at circles like these. Here theory and practice came together, reason was confronted with experiment in a self-conscious manner, and applied to specific technical problems.

If astronomy was eclipsed in Italy after the condemnation of Galileo, this was hardly true of applied mechanics or natural history. Geographically and geologically, Italy was very ill placed to take much part in the great economic expansion of the late seventeenth and the eighteenth centuries. An air of decline hung over even the most flourishing parts of the country. Venice, Milan and Florence already looked as if they had known better days. However, if we should add up the contributions made by Italians during this age of decay, they remain quite substantial, not only in expanding branches of physics, such as hydrodynamics and electricity, but even more in natural history, in biology and geology. This paper has touched on a corner only of the scientific work done in late Cinquecento Italy; I hope it has suggested something of the fertility of the ground from which so fine a harvest was to grow.

References

1. R. Dallington (1605), *Survey of the Grand Duke's State of Tuscany in the year of Our Lord 1596*, London, p. 35. But on p. 31 he gives a list of 24 kinds of fruit, 22 kinds of vegetables and 18 kinds of grain grown in the region.

2. Prominent among many military engineers of the sixteenth–early seventeenth centuries are Turriano, Marchi, Campi and Targone. The idea of a career open to *artistic* talent is a commonplace of Renaissance historiography.

3. J. R. Hale (1965), "The development of the bastion, 1440–1539, an Italian chronology". In *Europe in the late Middle Ages* (eds. J. R. Hale *et al.*), London, pp. 466–94.

4. P. Ramus (1567), *Prooemium Mathematicum*, Paris, p. 469.

5. F. Marchi (1864), *Cento Lettere del Capitano Francesco Marchi Bolognese* (ed. A. Ronchini), Parma, p. 165. On the issue raised in this introductory section, the most enterprising guide remains the classic work of F. Braudel, *Civilisation Matérielle et Capitalisme*, Paris, 1967, and his *La Mediterranée et le Monde Mediterranéen à l'Epoque de Philippe II*, Paris, 1947. These of course deal with far wider horizons in time and in space: on the Venetian Republic a more detailed treatment may be found in *Crisis and Change in the Venetian Economy in the 16th and 17th Centuries* (ed. B. Pullan), particularly in S. J. Woolf, "Venice and the Terrafirma: Problems of the Change from Commercial to Landed Activities", pp. 175–203. On the relations of hydraulic engineering to theoretical fluid mechanics: H. Rouse and S. Ince (1963), *History of Hydraulics*, New York, pp. 59–63, 68–70. A. G. Keller, "Renaissance Waterworks and Hydromechanics", *Endeavour*, **xxv** (1966), 141–5.

6. S. Drake and I. E. Drabkin (trans. and annot.) (1969), *Mechanics in Sixteenth-Century Italy*, Univ. of Wisconsin Press, Madison and London; a very useful introduction to the subject of this paper, even if one may disagree with some of its assertions and regret its brevity. It includes translations of excerpts from the Quesiti, as well as from the Italian version of Guidobaldo's Mechanica. I have made use of these when translating passages from these works, while reserving liberty to amend where I felt it desirable.

7. On Tartaglia, the most convenient short account is facsimile edn. of the 1554 edn. of *Quesiti et Inventioni Diverse* (ed. A. Masotti), Brescia, 1959. Cf. also L. Bittanti (1871), *Di Niccoló Tartaglia, Matematico Bresciano*, Brescia.

8. N. Tartaglia (1554), *Quesiti et Inventioni Diverse*, Venice, pp. 69–70.

9. *Inventione nuovamente trovata da Niccoló Tartaglia* (1537), *Utilissima per Ciascuno Speculativo Matematico Bombardiero et altri . . . intitolata Scientia Nova*, Venice.

10. Drake and Drabkin, ref. 5, p. 18–19.

11. N. Tartaglia trans. (1543, 1565), *Euclide Megarense Philosopho Solo Introduttore delle Scientie Mathematiche*, Venice.
12. N. Tartaglia, *Quesiti* . . . f. 78 r.
13. On the history of the Mechanical Problems in the sixteenth century, two admirable studies have appeared in recent years: M. Schramm "The Mechanical Problems of the Corpus Aristotelicum, the Elementa Iordani super Demonstrationem Ponderum, and the Mechanics of the Sixteenth Century", pp. 151–63 in A. C. Crombie, J.D. North and M. Schramm (1967), "Physics and Astronomy", in *Atti del Primo Convegno Internazionale di Recognizione delle Fonti per la Storia della Scienza Italiana: I Secoli XIV–XVI* (ed. C. Maccagni), Florence; and P. L. Rose and S. Drake (1971), "The Pseudo-Aristotelian Questions on Mechanics in Renaissance Culture", *Studies in the Renaissance*, **xviii**, pp. 65–104.
14. A very brief account of his life and work, with bibliography, may be found in P. L. Rose (1971), "Materials for a Scientific Biography of Guidobaldo del Monte" in XIIe *Congrès International d' Histoire des Sciences Paris 1968, Actes xii* (Monographies de Savants), Paris, pp. 689–772. G. Arrighi (1968), "Un Grande Scienziato Italiano, Guidobaldo dal Monte", *Atti dell'Accademia Lucchese di Scienze Lettere ed Arti, Nuova Serie*, **xii**, contains a number of letters chiefly relating to the nova of 1604; I have been unable to consult A. Favaro (1899), "Due Lettere Inedite di Guidobaldo del Monte a G. Contarini", *Atti del Reale Instituto Veneto di Lettere Scienze ed Arti*, **lxix**, to which Dr Rose makes reference.
15. Guidobaldo de'Marchesi dal Monte (1577), Mechanicorum liber, Pesaro; cf. S. Drake and I. E. Drabkin (ref. 5), pp. 239–328; Guidobaldo's introduction, pp. 241–7.
16. For his biography: E. Salaris (1913), *I Savorgnano*, Rome, p. 67 ff.
17. N. Tartaglia, *Quesiti* . . . f. 34 r.
18. N. Tartaglia (1551), *Regola Generale da Sulevare con Ragione e Misura non Solamente una Affondata Nave . . . intitolata la Travagliata Inventione*, Venice; Il Secondo Ragionamento. On Tartaglia's suggestion to use hydrostatic caissons for salvage work, and the attempt to put this idea into action in 1560, which offers another case germane to the theme of this paper, A. G. Keller (1971), "Archimedean Hydrostatic theorems and Salvage Operations in 16th-century Venice", *Technology and Culture*, **xii**, pp. 612–17.
19. F. Pigafetta trans. (1581), Guidobaldo de'Marchesi dal Monte, *Le Mechaniche*, Venice.
20. S. Drake and I. E. Drabkin, ref. 5, pp. 248–55. The figure of 14 000 lb mentioned below would seem to be an accepted weight for a cannon: cf. L. Collado (1592), *Platica Manual de Artilleria*, Milan, p. 73, who speaks of the great capacity of a screw-jack. It is not easy to see how it could be supposed that only *one ounce* of force was exerted.
21. J. Dee, Mathematical Preface to H. Billingsley trans. (1570), Euclid, *The Elements of Geometrie*, London, sig. d. ii; B. Lorini (1596), *Delle Fortificationi*, Venice, pp. 187–9.
22. The literature of mechanical invention took this topic up enthusiastically, for example J. Besson, *Livre Premier des Instruments Mathematiques et Méchaniques* (n.p., n.d.), plate 31; and the topic is discussed in many contemporary books on the art of war and the management of artillery, for example D. Ufano (1612), *Tratado de la Artileria*, Brussels, p. 142; G. Busca (1585), *Dell' Espugnatione et Difesa delle Fortezze*, Turin, p. 25; E. Gentilini (1598), *Instruttione di Artiglieria*, Venice, p. 35; B. Lorini, ref. 21; L. Collado, ref. 20.
23. Biblioteca Ambrosiana, R.121.sup., ff. 4–24. Cf. A. Rivolta (1933), *Catalogo dei Codici Pinelliani dell' Ambrosiana*, Milan, p. 119.

E. M. CUDAHY LOYOLA UNIVERSITY MEMORIAL LIBRARY

24. Ref. 23 ff. 4–6.
25. Ref. 23 ff. 9–10. It will be noted that Savorgnan was aware of the strain put on key elements of his engine, even if incapable of quantifying this strain, or of devising a means to countervail it.
26. Ref. 23 ff. 11–12.
27. Ref. 23 ff. 18–22.
28. Ref. 23 ff. 14–15; this and the answering letter are out of chronological order.
29. Ref. 23 ff. 17–19.

3. Science in the Italian Universities in the Sixteenth and Early Seventeenth Centuries*

C. B. SCHMITT

(*Warburg Institute, University of London*)

I Introduction

When one reads an account of early modern science in any of the numerous books now devoted to a treatment of that subject one cannot help but be struck by the way in which the universities are nearly always seen as holding back scientific advance in one way or another. The emergence of "modern science" is usually taken to be the results of throwing off the shackles of an old-fashioned Aristotelian worldview and its gradual replacement in the course of the sixteenth and seventeenth centuries by a "new" science. The roots of this new movement, so the story goes, can already be seen clearly with Copernicus and Vesalius, but it was only with Galileo, Descartes, Kepler, Harvey and Newton that it became consolidated and established firmly. Even though there was this forward thrust leading to the modern world, so the interpretation continues, there still co-existed alongside it the outmoded and superseded university culture of the Middle Ages, which had outlived its usefulness, but refused to die until long after it had served its purpose.

There is in this, as in nearly all historical generalisations, a grain of truth. There can be no doubt that the university culture of the sixteenth and seventeenth centuries was in many ways retrograde. On the other hand, perhaps both the intellectual level and the receptivity to new ideas found in the Renaissance university tradition was more progressive than is often admitted. The important contributions of Renaissance universities have been noted many times in a variety of specialised studies, often hidden away in publications of some local society. Consequently, much of the important material which has been collected over the past half century and more has not yet surfaced in the general surveys. Yet, when brought together, this

---

* The material contained in this paper was used as the basis for seminars given at Imperial College, London, and at the University of Kent, Canterbury.

35

material indicates that perhaps more significant scientific developments took place within universities than is generally realised.

For the present I shall focus upon the teaching of scientific subjects and, to some degree, also the research which took place in Italian universities during the century and a half separating the arrival of the young Copernicus there and the death of Galileo. I plan to give a survey of the sort of scientific thought which prospered in the universities, paying particular attention to several specific changes and new developments found in that period.

As an introduction to this subject it might be well to make some general remarks concerning the structure and orientation of the Italian universities of the Renaissance. While they shared many characteristics with those of other countries, Italian universities also had certain peculiarities of their own. It is important to emphasise this, for most general accounts of medieval universities focus unduly upon those of northern Europe and some of the significant aspects of Italian university education are passed over in silence.

Leaving aside points of individual structural variation, we can say that for our purposes the major differences between Italian universities and the trans-alpine ones was the organisation of the curriculum. In the north liberal arts and theology were strongly emphasised, while in Italy the bias was more towards the professions of law and medicine. Like all generalisations, this one has many exceptions. By and large, however, it holds and many examples could be cited of Italian students going to Paris or Oxford to study theology or of northern Europeans going to Italy for medical or legal studies. Moreover, arts faculties did not develop in the same way in Italy as they did in Paris or in the German universities, for example. What liberal arts subjects there were in Italy were principally directed towards preparation for medical or legal studies and, for the most part, the arts degree was not looked on as an end in itself, as it was in Paris as early as the thirteenth century. Thus, with few exceptions, a broad liberal education was not the aim of Italian universities of the Middle Ages and Renaissance.[1] As evidence of this we can note that Aristotle's works on moral philosophy were read only infrequently as part of university courses in Italy, but were much more integral to the arts curriculum of the north.

The general pattern of university structure which we find in Italy at the beginning of the Renaissance is one of a three-faculty university: law, medicine and theology. The first of these seems to have been the most important, both in terms of the prestige and remuneration of the law teachers and in terms of student numbers, where approximately two-thirds of those enrolled in the universities at any one time were law students.[2] At the other end of the scale was theology, very much a splinter subject with a very small teaching staff and relatively few students. Though small, the faculties of theology were important and often, as at Padua, were tied to other extra-university educational institutions run by various religious orders.[3] With regard to the subject of theology in Italian universities we must maintain a balanced perspective. At

Padua, for instance, the theology faculty was relatively small, though it still played a significant role. Padua was not the anti-theological and secular institution which it has sometimes been considered to be.[4] Before turning to medicine, let us once again underline the facts that (1) in terms of prestige and size, the faculty of law was of the greatest significance by far and (2), though small, there was nevertheless a faculty of theology in most universities and, in some cases, it was quite influential.

The centre of our focus must be what I have termed the faculty of medicine. Terminology in this matter was not fixed and the relevant university faculty and the degrees coming from it went by a variety of other names, including: arts, arts and medicine, philosophy and medicine. The education gained in this faculty contained a significant component of arts subjects, particularly philosophy, but the goal of study was predominantly medical. Though it was possible for a student to take a degree in "arts" or "philosophy", this was not usual. Most students continued until they took a degree in medicine, sometimes also called "in arts and medicine" or "in philosophy and medicine". I hope that all of this is not too confusing as it is valuable to get this matter sorted out before proceeding and to realise that by whatever name the faculty was called, the predominant emphasis was on medical subjects, though logic, natural philosophy, mathematics, or botany may also have been studied along the way. What I have to say in the remainder of this paper relates to the medical or arts faculty, for that was where such scientific interests as there were were pursued.

Medical education, as it was organised in the Italian universities of the Renaissance, was divided into a two-tier curriculum which covered a normal five-year course of studies.[5] During the first two years the student was given a normal grounding in logic and natural philosophy and after that he progressed to medical studies themselves. Logic and natural philosophy courses were based on Aristotle, those parts considered to form a beneficial foundation for medical studies being emphasised. In logic, the *Posterior Analytics*, Aristotle's work devoted to setting forth a scientific method, was emphasised and in natural philosophy the mainstays were works such as the *Physics*, *De anima*, *De generatione et corruptione*, and *Parva naturalia*. Medical studies were, in turn, divided into two parts, theoretical and practical medicine. These two branches were studied simultaneously in a three-year cycle of courses. The entire curriculum was based on Hippocrates, Galen, and the Arabic physicians Avicenna and Rhazes. In addition to these subjects, anatomy and surgery were also imparted to the students, increasingly, it seems, as separate subjects rather than as part of theoretical and practical medicine. In the course of the sixteenth century a separate chair of botany was also introduced into most universities and this subject grew in importance in a way we shall detail later. University education in arts also included, from time to time, some mathe-matics, humanities, moral philosophy and occasionally other subjects, but all of these occupied a very subsidiary position in the overall curriculum. For the purposes of the present paper, mathematics teaching is of some importance and we shall have more to say about that later. It should be noted, however, that the subject was a relatively minor component of the overall curriculum.

This, then, in brief summary is how Italian universities of the Renaissance were organised. The basic orientation was towards the practical professions of medicine and law. What scientific studies there were, were cultivated almost exclusively as adjuncts to medical education. The basic component of the scientific education was most traditional with instruction being based largely upon Greek and Arabic materials, in Latin translation of course. As we shall see there were certain exceptions to these generalisations and, for the purposes of the present paper, the exceptions are perhaps more interesting, for they show tendencies towards innovation and change.

The structure of such a curriculum and the emphases on its different sections, as well as the range of materials omitted from it might startle those used to the university teaching of the nineteenth and twentieth centuries. It was, indeed, a very different world from ours. Here, I should like to raise a question which, strictly speaking, is not wholly appropriate to the data at hand. What were the scientific subjects and how did they function in the overall curriculum? Strictly speaking, the "scientific" subjects included those which from our point of view scarcely seem to have a scientific component at all, namely the Aristotelian writings on natural philosophy and logic, which formed the core of the curriculum. However, looking at it from an historical, but more modern, viewpoint we can note that there was included in university instruction a variety of subjects which ultimately proved to be fruitful from the point of view of modern scientific developments and in retrospect can be seen as pointing in the direction of "scientific progress". Therefore, a selective process can draw out a number of elements of sixteenth and early seventeenth-century Italian university culture which were not only not so backward-looking as is normally thought, but perhaps even represent some of the more progressive tendencies. Thus articulated, the "scientific" (those aspects which materially contributed to the development of those disciplines which came to be called "scientific" in more recent times) component of Italian Renaissance universities can be isolated and treated in a way to bring them into accord with more recent views of what "science" is and with what "history of science" should concern itself. In fact, once it is realised how very different was the orientation of universities of the Renaissance from those closer to the present day, university contributions to scientific knowledge are found to be by no means small.

Perhaps even more than elsewhere it was in the general field of medicine and in the more restricted and peripheral one of mathematics that interesting things were happening in the Italian universities. Medicine, for one reason or another, has remained primarily a university subject down to our own day.

While revolutionary work in mechanics, pneumatics, or astronomy were done in non-university environments during the early modern period, nearly all of the advances in medicine were made within the university framework. Though extremely conservative in some respects, medical men also adapted themselves to change and innovation when the need arose and the extent of the innovation and improvement which took place within university medical studies between 1500 and say 1650 was quite considerable, especially when compared with the situation in fields such as physics and psychology. The significant advance in mathematics during the period has been known for a long time, but what is not generally realised is the richness of the university mathematical culture which parallels that which developed in other cultural centres. When one considers the range of mathematical work carried on by university men during the period being considered, one finds that it goes far beyond the rather meagre provisions of the statutes.

With these preliminary comments in mind, let us now look into some of these matters in greater detail, paying particular attention to the development of scientific ideas within medicine and mathematics. We shall first turn to medicine.

As noted above, medicine constituted a large portion of the arts faculty and in the larger universities of Italy was taught by a substantial staff numbering between ten and twenty.[6] Medicine was indeed the primary purpose for the existence of the arts faculties; logic and philosophy were mere propaedeutic studies to give students the necessary background for medical subjects. In numerous ways medicine represents the most progressive and vital scientific strand of Italian Renaissance universities.

Recent historiography of science has tended to see in the development of physical sciences the model by which other intellectual movements are to be judged. Nevertheless, there is an equally good reason to see medical and biological sciences as an important facet of the development of modern civilisation and one which has its own internal logic not necessarily reducible to physical science. Unlike the physical sciences, the bio-medical ones advanced nearly wholly within a university context. In fact, when the medical and biological sciences are seen as equal partners with the physical sciences in the "scientific revolution", not only universities, but also the Aristotelians who flourished within them appear in a somewhat different light. From being the bogey men who held back progress in the natural sciences, they become important contributors to the emergence of early modern science.[7]

II   Botany

In botany I think we have one of the most important points of university scientific development of Renaissance Italy.[8] This was the science *par excellence* which emerged from oblivion during the sixteenth century to become a growth point from which a number of other natural history subjects developed. The relation of botany to medicine had been firmly established in Antiquity and the bonds between the two had been even more closely cemented during the Middle Ages. It remained, however, for the sixteenth century to develop botany as a fully autonomous subject in the medical curriculum. The first separate chair of botany was apparently established at Rome in 1513, but it seems to have been rather short-lived and did not exert any significant influence on the course of future events. A few years later we find other botanical chairs being founded with greater effect. These include Padua (1533), Bologna (1534), Ferrara (1543) and Pisa (1544). Later the teaching of the subject spread to other universities, including Naples, Messina and Rome.

At the fountainhead of botany instruction and research was Luca Ghini (*ca.* 1490–1556),[9] who was one of those persons fortunate to be at the right place at the right time. Just as botany was entering the university curriculum in Italy Ghini was in a position to benefit from the new enthusiasm and support for the discipline. He was the first lecturer in the subject at Bologna in 1535 and then moved to Pisa in 1544, where once again he put the study of the subject on a firm foundation. Among his students are to be numbered Aldrovandi, Anguillara, Cesalpino and Maranta. Unfortunately, we have very few of Ghini's own writings and what we know of him and his work comes largely from the testimony of students and other contemporaries. Nevertheless, the information which we can gather on him shows most clearly his position as an innovator in the newly founded university subject.

Ghini's own teaching career parallels the development of botanical studies in Italian universities. He was first appointed to teach medicine, but after a few years he was assigned to teach botany on a year by year basis. In 1539, he was given a five year appointment as a botany lecturer and after that time the subject was firmly entrenched in the Bologna medical faculty. By then we can already note that botany teaching was envisioned to encompass a rather broad range of responsibilities. According to a contemporary university document Ghini's duties were ". . . *simplicis medicinae tradendae et monstrandae munus, eamque ex Galeni et aliorum medicorum veterum et recentiorum libris in publicis scholis interpretandum.*"[10] The term *monstrandae* is important here. It indicates the practical nature of botany instruction and the fact that it was part of the task of the lecturer to show the students the actual plants which were being discussed. A similar document from Pisa a few years later is even more specific in this respect, detailing that the botany lecturer was to "ostendet plantas".[11] According to his student, Luigi Anguillara, Ghini "*faceva esperienze sopra delle piante*".[12] What precisely this means is difficult to assess, but at the very least it means that he worked with plants themselves and his theoretical knowledge was corroborated by a familiarity with the materials being studied, which perhaps could not be said of most lecturers on physics (or natural philosophy) of the time. The range of Ghini's interests and his teaching activities, which extended far beyond the bounds of botany are alluded to by another of his students, Benedetto Varchi who later achieved fame as a historian and literary figure. Varchi indicates that in addition to botany Ghini also lectured on minerals and professed an interest in alchemy.[13]

Though Ghini's direct influence through publication was practically nil, the impact he had on the development of the new subject was quite considerable. The broad influence of botanical studies later in the century is perhaps epitomised by Ulisse Aldrovandi (1522–1605)[14] better than by anyone else, but before turning to him as a representative of the situation a few decades after Ghini's death, let us look at some other aspects of sixteenth-century botany. One of the most important new developments in university science of the sixteenth century was the establishment of botanical gardens.[15] Along with the development of the public anatomical theatre, the university botanical garden probably represents as well as anything else the new sort of material environment which was becoming a part of scientific study. The first university botanical gardens were established at Padua and Pisa, both about 1544.[16] During the next decades other universities followed the model of the innovators and by the early seventeenth century a botanical garden was considered part of the necessary physical trappings of any self-respecting university. In fact, botanical gardens were taken quite seriously. Pisa's was admired by a multitude of visitors, including Pierre Belon, Pierre Pena and Mattias Lobel[17] and when Pope Pius V decided to found his own garden at Rome, he sent a representative to consult with Cesalpino for expert advice.[18] The prestige which university gardens had acquired by the seventeenth century can be seen in the elaborate designs and careful attention to aesthetic detail so apparent in the plans of the Messina garden laid out by Pietro Castelli.[19] As already mentioned, however, the gardens were far more than show places and served as an important teaching instrument in the training of physicians.

The number of gardens continued to grow and many newly discovered species were naturalised into them, but there was even more. Already in the middle of the sixteenth century we find that specific field trips were being made, not only privately, but as extensions of university courses in botany. For example, in the section of Aldrovandi's autobiography concerning the year 1557 we read the following

> "Having obtained a licence from the Senate, in May he [i.e. Aldrovandi] went forth with many students, first seeing all of the valleys from Padusa as far as Ravenna, in which he observed the major part of the plants [described] by the ancients as well as many which they did not describe. Then he left Ravenna for Rimini, from Rimini to Avernia, where St Francis lived, and there he found many plants not described by the ancients. Having returned from Rimini, he saw the garden of Guilio Moderati, made famous by Mattioli . . ."[20]

In fact, it seems as though field trips were not so unusual as we might expect at that time, for we also have a letter of Pier Andrea Mattioli to Aldrovandi in which he describes a similar trip.[21]

Though the institution of botany as a university subject was originally meant to impart to students information on the plant world and the use of herbs for medicinal purposes, it did not take long for the subject to acquire broader implications. This is best exemplified in the case of Aldrovandi, of whom we shall speak in more detail below. It was not only with him, however, that we see university botany lectureships blossoming forth into a sort of general natural history subject. In addition to the plant world, the animal and mineral worlds were often included in the teaching and research interests of the botanists.

Concurrent with the branching out of botany into other fields we see the establishment of various repositories meant to preserve the natural knowledge which was being accumulated. Herbaria were of course formed as adjuncts to the gardens,[22] but natural history museums of a more general nature were not uncommon. That of Aldrovandi is perhaps the most distinguished. Part of the collection is still extant and can be seen at the University of Bologna.[23] In its time it was important as a study collection, which steadily grew in size during Aldrovandi's lifetime. In it were to be found an extensive range of specimens from the plant, animal and mineral kingdoms. By 1570 it had already reached an impressive size, among other things, containing fourteen volumes of specimens, 4500 minerals and gems and many paintings of animals. After Aldrovandi's death, a new six-room addition was built onto the Palazzo comunale in Bologna, where his collection was put into a public museum where it remained for the instruction and delectation of the Bolognese for a century.

As botany progressed in the sixteenth century, both inside and outside of the universities, it developed into a science embracing what was perhaps the first real international community of scientists of modern times. Inevitably there were petty disputes, jealousies, and disagreements, but in looking over the situation as a whole one cannot help being impressed by the degree to which early modern botanical studies were marked by an altruism and free interchange of scientific ideas. There still remains much evidence of the epistolary interchange of information and specimens among botanists throughout Europe. As yet this material has been imperfectly studied and much of the most significant evidence lies unpublished in many libraries. The publication of the Gesner botanical drawings with his comments from manuscripts preserved in Erlangen discloses well the wide range of sources from Italy to Britain from which he procured his seeds, specimens and information.[24] The same thing appears in the vast number of letters from Daléchamps in Lyon, Camerarius in Nuremberg, Aldrovandi in Bologna and numerous others. In all of these we see a dedicated determination to advance scientific knowledge and to exchange ideas and botanical specimens freely, with there being little evidence of envy, of competing interests or of national or religious differences. Many examples could be cited, but the spirit is adequately typified by a letter sent from Joachim Camerarius, a German Lutheran, to Ulisse Aldrovandi, an Italian Catholic, dated 1579: "I have metallic and fossil specimens and daily many are being sent to me by friends; if I can know which of these are still lacking to you, I shall gladly send them to you. From you, in turn, I ask that you also help my study by (sending) rare seeds and plants which I truly need."[25] A list of the plants and seeds which Camerarius would have liked was appended to the letter.

The vast botanical compilations of the sixteenth century, regardless of their shortcomings in terms of classification and of accurate description, became progressively more comprehensive in the number of plants they listed and it was in no small measure due to this free interchange of information. In fact, I think that we can see among these botanists—as perhaps also among the classical scholars of the same period—the forerunners of the seventeenth-century *respublica literaria*, which played such an important role in disseminating information in the years before the establishment of scientific journals and periodicals.

Perhaps no one in Italy was more important than Aldrovandi in giving botany an important place in university education and in making it an academic subject of broad scope and of international significance. After having studied law as a young man he then moved into medicine and came under the influence of Luca Ghini, who inspired him to an interest in botanical studies. His teaching career began about 1553 when he was made a lecturer in logic and philosophy, but in 1556 he began to teach botany and continued to do so until his death in 1605. Among other things, he was perhaps the first in Italy to lecture on Theophrastus' botanical writings in a university course.[26] About 1560 the name of the chair was changed from "botany" to *lectura de fossilibus, plantis, et animalibus*.[27] This title means what it says, as we find from a letter to Aldrovandi from Maranta in Naples who says: "I am pleased to learn of your honour of being elected to the lectureship in natural philosophy, plants, animals and fossils, in which you teach now one and now the other in alternating fashion, for such lectures are certainly of great importance."[28] Of the implications of this we have already hinted, but it should once more be underlined that here we see the beginnings of the teaching of the biological sciences as we understand them. Aldrovandi was not tied to a specific teaching duty, but could lecture on a wide range of natural history subjects to his students. The instruction was inevitably based not only on Aldrovandi's wide experience in the field, but also on his immense collection of natural history specimens. As such, he was perhaps the first great naturalist to give to a university the benefits of a broad range of capabilities. The credit for this goes as much to the University of Bologna as to Aldrovandi. It had been late in founding a botanical garden (1568),[29] but it gave more or less free rein to a man who could teach students much which fell outside the normal curriculum. This tendency in Italian botany teaching can be traced into the seventeenth century, when, for example, Pietro Castelli in establishing a garden at Messina insisted upon the same extensive range of concerns in the interest of medical practice.[30]

Without lingering too long on the subject of botany, let us summarise some of the contributions made by the subject as it was taught in the Italian universities of the Renaissance. First, it showed itself to be an academic subject capable of expanding from a relatively narrow intellectual basis into a sort of catch-all natural history subject. Secondly, unlike the physical sciences it was wholly rooted in the collection of a wide range of empirical data. Thus, even if the theoretical structure of the science later proved to be unviable, the discrete observational data were still scientific facts which could be utilisable within a different framework. Thirdly, it allowed a balance between practical usefulness in teaching and medical practice on the one hand and the accumulation of new information based on observation and research, on the other. Fourthly, in its position as a relatively comprehensive natural history subject, it could dovetail easily into schemes making use of the expanding world of iatrochemistry and alchemy. Botany thus had an adaptability which made it a central subject as medical and scientific education progressed in seventeenth and eighteenth-century universities.

While we have perhaps unduly emphasised botany, it should be noted that within the framework of medical studies it was not the only important development to come from the Italian Renaissance universities. The remarkable unfolding of anatomical science in the course of the sixteenth century is well known.[31] Plastic surgery originating with Tagliacozzo[32] and the seventeenth-century work of Borelli[33] and Malpighi[34] were also fostered within the Italian universities. Medical education of a practical nature was enhanced by direct medical experience with patients afforded by Paduan medical education, which made several provisions for that aspect of the training.[35] Indeed much more could be said of these matters, but let us now turn to a different aspect of science in universities—mathematics.

III   Mathematical Subjects

As already alluded to, mathematical studies occupied only a very peripheral position in the curriculum. If a dozen or more medical men were teaching at one of the larger universities at any given time, the usual complement of mathematicians was but one. Rarely two or three lecturers were engaged in teaching the subject simultaneously, but that was somewhat unusual. Mathematical studies were not considered very significant in the overall educational scheme. In the Italian universities, at least, this subject was cultivated primarily as an adjunct to medical education, though, as almost always happens, we find a few in the field who began to study the subject for its own sake or as it applies to other matters little related to the original intention when instituted.

In general, we can say that the mathematics taught in Italian universities of the sixteenth century encompassed a somewhat broader field than our word "mathematics" suggests. The usual teaching texts were Euclid's *Elements*, Sacrobosco's *Sphaera* and various writings of Ptolemy. As we shall see, a wide variety of other things were also taught under the same rubric, but even a somewhat conventional curriculum included astronomy, astrology and geographical studies, in addition to mathematics. In fact, when we look at the actual university records we find the terms "astrology", "astronomy", and "mathematics" used indifferently to describe the teaching activities of the incumbents in these positions.

An important fact to keep in mind is that as "mathematics" developed in the Renaissance it encompassed two quite different and, to us, inimical traditions. In addition to the technical mathematical tradition as we normally understand it, there was also an equally strong—or, perhaps, even stronger— tradition of the darker side of the mathematical arts. While from the modern point of view, much which was cultivated along these lines seems not only of little interest to the historian of science, but perhaps even irrelevant to the enquiry, it must be noted that the two traditions of mathematical arts were inextricably entwined. Consequently, it is often difficult to see where one side ends and the other begins. Even more difficult is any sort of objective evaluation of the merits of mathematical and astronomical works in which the two aspects are so intimately interwoven. The problem of isolating the two strands and evaluating them separately is evident not only in critical studies on Girolamo Cardano and John Dee but even on Kepler.

The mystical and pseudo-scientific side of Renaissance mathematical arts was not merely the predilection of the private scholar, but was also most evident in university mathematics. Indeed, even a cursory glance of the mathematics taught in various Italian universities during the Renaissance reveals that the same twofold tendency can be discerned there. Which of the two gained the ascendency at any particular time or place may have something to do with local conditions, but as yet I have been unable to determine to what extent this is so. What does emerge, however, is that the mathematical tradition which Galileo inherited at Pisa[36] was significantly more closely attached to the mystical and astrological tendencies than was that at Padua, which he inherited upon becoming mathematics lecturer there in 1592.[37] This is not to say that *only* astrological mathematics was taught at Pisa before his time and *only* technical mathematics at Padua, for there were elements of both at both places: the dominating tendency was different at the two places, however.

Here we are primarily concerned to trace the more positive contributions of universities, so we shall concern ourselves mainly with mathematical and scientific subjects in the conventional sense which were practised there. There can be little doubt that certain aspects of the more extravagant mathematical and astrological speculations contributed to early modern scientific developments. These must be bracketed as far as possible for the present purposes, however, if we intend to illustrate some of the more positive contributions of mathematics within the universities.

As already noted "mathematics" as the term was understood in the sixteenth-century university curricula covered a fairly wide range of mathematical and scientific subjects. In addition to basic geometry and Ptolemaic astronomy, other subjects which were taught from time to time and place to place include optics, mechanics, geography, cosmography (a combination of astronomy and geography), anemography and hydrography. While it is unlikely that any student of the sixteenth century could have heard lectures on all of these subjects during his university years, it would not have been unusual for him to have made contact with several of them. Geography and cosmography were quite widely taught, principally from the Ptolemaic basis which the statutes allowed. Nevertheless, a goodly amount of new material was added, including some new information derived from the recent geographical explorations. One of the new types of works which seem to have emerged in Italian university teaching of the time were *compendia* of geography and astronomy, a hybrid subject which sometimes went under the name of cosmography.[38] For example, Galileo's predecessors in the Pisa mathematics lectureship, Giuliano Ristori and Filippo Fantoni, had a hand in the composition of such a work.[39] Moreover, the term *cosmographia* begins to appear in the official title of the lectureship in mathematical subjects at Ferrara about the middle of the sixteenth century, thus indicating that the subject had become an integral part of the mathematical arts by that time.[40] Mechanics taught from the pseudo-Aristotelian work of that title, was lectured on quite frequently at Padua between 1548 and 1610 by Catena, Moleto and Galileo.[41] Optics was also taught there several times from Euclid's treatise on that subject, thus introducing a new subject within a traditional framework.[42] From Antonio Riccoboni, the first historian of the University of Padua who himself lectured on humanities there in the late sixteenth century, we learn that Giuseppe Moleto also taught what is termed "anemography and hydrography."[43] Precisely what this can mean we can only surmise. Such a subject could be rooted in Ptolemy, growing out of problems raised in the *Geography*, perhaps dealing with winds and rain as they relate to geographical studies, but perhaps also related to Aristotle's *Meteorology*.[44]

Some of these things can still be studied more deeply and, when this is done, will undoubtedly yield additional information on the range of mathematical subjects taught in sixteenth-century universities. As yet the extant mathematical texts, both printed and manuscript, which came from the Renaissance university culture have been but little studied. From the range of subjects just described we can see that mathematics was a much richer field of study than one would imagine from reading the statutes of the various universities. Indeed, though there is no evidence for this in the statutes, we know that a sixteenth-century text such as Oronce Finé's *Sphaera*, first published in 1542, was used as a textbook at Pisa towards the end of the sixteenth century,[45] as it was also being used in various northern European universities.

In addition to extending into new fields of mathematical arts and expanding on the range of subjects taught we also find the mathematics lecturers concerning themselves with fields of enquiry which had previously been the province of others. Particularly fruitful, it seems to me, were the incursions of several mathematicians into the field of philosophy. Though an attempt was made in the fourteenth century to join mathematical analysis to investigations of natural philosophy, these studies seem to have lost their attraction during the fifteenth century. Yet, it was when the study of mathematics was rejoined to that of natural philosophy by Galileo that the so-called "scientific revolution" got off the ground. Remarkably little scholarly work has been done until the past few years to try to illuminate the mathematico–natural philosophical conjunction during the century before Galileo. Attention rather has been bestowed upon the methodological discussions of the sixteenth century within the framework of logic *per se*[46] and the rather frequent excursions of mathematicians into philosophy have been but little studied. Yet the new directions of research of several scholars during the past few years indicate that this may be a most fruitful avenue of approach through which to understand certain aspects of the roots of the scientific revolution.[47]

Here, I shall concentrate upon two principal lines of development: (1) interest of mathematicians in problems of natural philosophy, and (2) the interaction of mathematicians with problems of Aristotelian methodology and logic. Other important areas, such as the revivification of Platonic views on mathematics and the recovery of new texts, including Proclus' commentary on Euclid's *Elements* and the repercussions of this must be left for another occasion.[48]

We have several specific examples of mathematicians extending the range of their interests into problems specifically considered to fall within the province of natural philosophy at the time. For example, Federigo Delfino, a lecturer in mathematics and astronomy at Padua during the first half of the sixteenth century, wrote a work on tides which was published by the Venetian Academy after his death.[49] This, of course, furnished a precedent for Galileo's interest in the subject, as Moleto's unpublished treatise on fortification dating from 1575[50] foreshadowed Galileo's own work on the subject. Perhaps an even more pertinent example of a mathematician being interested in a problem of natural philosophy is furnished by Filippo Fantoni's treatise on motion.

Fantoni, as mentioned above, was Galileo's immediate predecessor as mathematics lecturer at Pisa and, indeed, was lecturing on the subject when

the young Galileo was a student there. Among the manuscripts which he left
behind is one entitled *Absolutissima quaestio de motu gravium et levium ... excerpta
in Accademia Pisana.*[51] This is a rather straightforward treatise, not much
different from conventional scholastic work on the subject such as that by
Zimara, Borro or Buonamici, to cite but three examples from the sixteenth
century. The content of Fantoni's question betrays no attempt to frame the
problem in terms of the Archimedean analysis which Galileo's *De motu* of a few
years later uses. Nevertheless, it is not without importance that a lecturer in
mathematics has taken it upon himself to treat this subject. It shows that
Galileo's own venture into the philosophy of motion was not without precedent
among the mathematicians who were his predecessors at Pisa. In fact, if the
title of Fantoni's treatise can be trusted, it would seem to indicate that he
lectured on the subject of motion at the university.

Interest in the Aristotelian background of Galileo has thus far centred
primarily on a possible link between the Renaissance methodological discussions regarding the *Posterior Analytics,* especially as the work was treated in the
universities of northern Italy. More important avenues of scholarly enquiry on
this point would seem to involve the attempts of the mathematicians and
philosophers to relate mathematical studies to Aristotelian views on scientific
methodology. There seem to be at least two different ways in which this
concern was manifested. Firstly, there appeared several works by mathematicians in which attempts were made to analyse and understand the various
Aristotelian texts which touch on mathematics.[52] Secondly, there was a dispute
on the certitude of mathematics which involved the basic question of whether
mathematical reasoning or syllogistic reasoning held the higher degree of
certitude.[53] This is not the place to go into a detailed discussion of these
matters, which call for further study. What these things show is that during
the sixteenth century in Italy we find a concern on the part of mathematicians
for some of the traditional problems inherent within Aristotelianism. Such an
interest could hardly fail to bring about a new approach to these questions
so long discussed by philosophers and logicians.

Many other aspects of university mathematics have been left unexplored.
For example, it would be worthwhile to look into the role of the men we have
discussed (and some others) in the debate over calendar reform which
attracted so much attention during the sixteenth century. Still, even our rather
brief treatment indicates that university mathematics instruction developed in
at least three important new directions. First, the range of the subject covered
under the umbrella of mathematics expanded quite significantly during the
sixteenth century, so that eventually quite a number of discrete scientific
subjects clustered around the mathematics chair in various universities.
Secondly, there was a certain degree of interaction between the problems of
logic and those of mathematics. Thirdly, there are a number of instances in
which mathematicians involved themselves with the traditional subject matter

of the natural philosopher. All of these indicate that there was a developing concern on the part of mathematicians to expand the function of their discipline. Out of these various interactions developed certain interdisciplinary characteristics which came to be a permanent feature of the intellectual map of early modern Europe.

IV   Conclusions

In concluding, I should like to try to summarise briefly what I take to be some of the significant changes which took place in Italian universities of the sixteenth and early seventeenth centuries. This will relate to universities in general, though it also relates to scientific studies in the artificially narrow sense in which we have defined them. We have already spoken at some length of mathematics and certain aspects of medical studies, but for a general picture of Italian universities other things must be taken into consideration. Even to understand the history of science these things are important, for, as we all know, the development of science is aided or retarded not only by the internal intellectual factors but by ones impinging upon it from outside as well. For the period we are considering perhaps two particular factors were important on university development in Italy: humanism and the Counter Reformation.[54]

We have already had occasion to allude to the impact of humanism on scientific studies. The new mathematical texts discovered, edited and translated by the humanists were no small factor in the university mathematical teaching which we have discussed.[55] Of perhaps equal importance was the recovery of other ancient scientific works unknown to the Middle Ages, for example Theophrastus' writings on botany, by far the most important writings on that subject before the sixteenth century. In the course of the period 1400 to 1650 these important writings, wholly unknown to the Middle Ages, were rediscovered, edited, translated and assimilated into the mainstream of European science.[56] In addition to the discovery of wholly new texts, however, the entire methodology of study developed by the humanist movement was an important force. For example, new and better texts of Aristotle were prepared and both Aristotle and medical authors were increasingly read in Greek. Moreover, though examples were not widespread, there are some instances of the teaching of philosophical and medical subjects from the original Greek texts rather than through the distorting veil of translation. A special chair for the study of Aristotle in Greek was established at Padua in 1497[57] and one for Hippocrates was instituted at Ferrara in 1562.[58]

Besides recovering important scientific and mathematical works from antiquity the humanists also made accessible varied materials illustrating ancient alternatives to the Aristotelian natural philosophy. Perhaps the most important of these alternatives, at least for scientific thought, was atomism which had a great impact on seventeenth-century thought, particularly through Gassendi.[59] In sixteenth-century Italy, however, we find Platonism to have gained a foothold, even in conservative university circles.[60] Thus, at the end of the century, there were university chairs dedicated to the teaching of Platonic philosophy at Ferrara, Pisa and Rome. While such an espousal of Platonism within universities was of limited scope and influence and much less intense than the extra-university development, it does nevertheless show that peripatetic tendencies did not dominate the universities so completely as sometimes supposed. A Platonist at Pisa was Jacopo Mazzoni,[61] a close friend and correspondent of Galileo, and Francesco Patrizi,[62] who taught the subject at Ferrara and later at Rome, can be directly linked to Newton through both Henry More and Gassendi, though in radically different ways. Consequently, Platonically-oriented thinkers did begin to find their way into university teaching, thus allowing students to come into close contact with philosophical ideas other than those of the peripatetic tradition.

The hardening of religious attitudes which the Counter Reformation brought into Italy was of serious consequence so far as the intellectual life of the universities was concerned. The effect on the development of science and philosophy was quite marked, though perhaps not quite in the way often supposed. There were certainly repressions and lack of freedom at some times and some places, but I seriously doubt whether it was quite so devastating to intellectual life as nineteenth-century liberal and protestant historians have made out. More important than the active repression of new ideas was the fact that theological and moral issues increasingly gained the attention of intellectuals and various clerical and religious activities became increasingly attractive professional options. The situation would seem to be akin to that of the late 1950s and 1960s in which the humanities were not quite repressed *per se*, but the material rewards were far greater for those who chose scientific studies.

Before making the next point, it would be well to dispose of two hoary stereotypes about Italian life of the period. First, before the Council of Trent, theology played very little role in Italian university life. Most universities did have faculties of theology, it is true, but they were quite small and insignificant compared to faculties of law and medicine. In no way could theology training in Italy compare to that of northern universities and it must be remembered that both Thomas Aquinas and Bonaventura went north for their theological studies. Moreover, all through the Renaissance the way to advancement in the Italian church was not through theology, but through law.[63] Secondly, though the Jesuits did gain some strength in Italy they did not take over the universities as they did in Catholic Germany and Eastern Europe. They founded the Collegio Romano and important schools at Messina and elsewhere, but had relatively little influence within the existing universities, their failure to get a foothold at Bologna being particularly well-documented.

If medieval Italian universities were not particularly theological in orientation, we see significant changes during the century after Trent. The already mentioned Jesuit Collegio Romano was founded with definite Counter Reformation purposes in mind. This is not to say that sciences and other secular subjects did not flourish there, but the general orientation of the College was towards theology and apologetics, science and mathematics being the icing on the cake as it were.[64] Perhaps an even more dramatic indication of the changing scene in Italy is the case of other universities, where more and more new teaching positions were established in theological and moral subjects after about 1550. This can be seen in various centres. In general, new chairs for the teaching of sacred scripture, Hebrew, theology, moral philosophy and allied subjects were instituted quite frequently during this period.

Bologna offers a particularly instructive example.[65] Before 1550 the university had one chair of theology which was not always occupied. Indeed in the year 1550 there was no theology taught. By 1580 there were three teaching positions in subjects related to theology, by 1600 there were six and by 1650 there were nine. Included in these numbers are several chairs of scholastic theology, the first of which was founded in 1588. Perhaps more than anything else this subject typifies the structured approach to theology encouraged by the post-Tridentine church. Examples could easily be multiplied: the first theology chair at Ferrara dates from 1569[66] and, at Pisa, the teaching of Hebrew and Sacred Scripture were added to the existing chair of theology in 1575 and 1589, respectively.[67]

Though Italian universities retained an important place as scientific institutions into the seventeenth century—the continuing medical reputation of Bologna, Padua and Naples clearly indicate this—the decline was already beginning. By the end of the century Italy had begun to fall behind Northern Europe, not only in science but in most other intellectual pursuits as well. In fact, in retrospect we see that Italy's pre-eminent position in science lasted longer than it did in classical philology, for example. Already by the middle of the sixteenth century the centre of gravity for classical studies had moved north. Thus Italy lost the grip she had had on a claim to intellectual excellence in the century during the Council of Trent. Galileo and others represent a late flowering of seeds which had been sown some time before.

References

1. One exception perhaps is Florence, which was not a typical university. On this see A. Gherardi (1881), *Statuti della Università e Studio Fiorentino*... Florence, and A. F. Verde (1973), *Lo Studio Fiorentino, 1473–1503: ricerche e documenti*, Florence (with quasi exhaustive bibliography on the subject in vol. **i**, 29–261).

C. B. SCHMITT

2. This estimate is based primarily upon the teaching staff at Bologna, Padua and Pisa and on the student numbers at Pisa. See the material cited below in ref. 6 and my "The Faculty of Arts at Pisa at the Time of Galileo" (1972), *Physis* **xiv**, 243–72.

3. See, *inter alia*, G. Brotto and G. Zonta (1922), *La facoltà teologica dell' Università di Padova. Parte I (secoli XIV e XV)*, Padua; the volume *Problemi e figuri della Scuola Scotista del Santo* (Padua, 1966 = Pubblicazioni della Provincia Patavina dei Frati Minori Conventuali, no. 5); L. Gargan (1971). *Lo studio teologico e la biblioteca dei Domenicani a Padova nel Tre e Quattrocento*, Padua.

4. In view of the good evidence brought forth in the publications cited in the preceding note, it seems difficult to maintain the position of Renan, Busson, Randall and others. On this see P. O. Kristeller (1968), "The Myth of Renaissance Atheism and the French Tradition of Free Thought", *Journal of the History of Philosophy* **vi**, 233–43.

5. I have tried to give a fuller analysis of Padua at the end of the fifteenth century in my "Thomas Linacre and Italy" (eds. F. Maddison and M. Pelling) (1974), *Linacre Studies*, Oxford (in press) and references to further literature will be found there.

6. As examples we might cite the following. During the second half of the sixteenth century there was an average of twenty-two medical teachers at Bologna, as can be learned from U. Dallari (1888–1924), *I rotuli dei lettori leggisti e artisti dello Studio Bolognese dal 1384 al 1799*, Bologna, ii. At Pisa there were nine in 1590 and at Padua there were eleven in 1592 as we learn from Galileo Galilei (1929–39), *Le opere... (ed. A. Favaro)*, Florence, **xix**, 39–42, 117–9.

7. More evidence for this point of view is given in my "Towards a Reassessment of Renaissance Aristotelianism" (1973), *History of Science* **xi**, 159–93.

8. There is much scattered information on this subject, but to the best of my knowledge there is no general study which makes it easily accessible. A useful general survey which contains much information is P. A. Saccardo (1895), *La botanica in Italia*, Venice (also in *Memorie del R. Istituto Veneto . . .*, **xxv**, no. 4).

9. On Ghini see especially G. B. De Toni's article (1921), in *Gli scienziati italiani* i (ed. A. Mieli), Rome, i, 1–4 and A. G. Keller's article in *Dictionary of Scientific Biography* **v**, 383–4 where further bibliography is given.

10. Cited in L. Sabbatani (1926), "La cattedra dei semplici fondata a Bologna da Luca Ghini", *Studi e memorie per la storia dell' Università di Bologna* **ix**, 13–53, 20, 32.

11. Cited in Schmitt (ref. 2), p. 254, n 57.

12. Sabbatani (ref. 10), 21.

13. B. Varchi (1827), *Questione sull' alchemia*, Florence, 34.

14. The best general survey of Aldrovandi remains the volume *Intorno alla vita e alle opere di U. Aldrovandi*, Bologna, 1907. Also useful, particularly for establishing the immense range of Aldrovandi's interests is L. Frati (1907), *Catalogo dei manoscritti di Ulisse Aldrovandi*, Bologna.

15. The history of botanic gardens has been but little treated in the scholarly literature. Particularly, are we in need of a good general treatment of the development of gardens in the Renaissance. Among the existing studies on the subject see A. W. Hill (1915), "History and Foundation of Botanic Gardens", *Annals of Missouri Botanic Gardens*, **ii**, 185–240; C. S. Gager (1937), "Botanic Gardens of the World: Materials for a History", *Brooklyn Botanic Garden Record*, **xxvi**; A. Chiarugi (1953), "Le date di fondazione dei primi orti botanici del mondo", *Nuovo giornale botanico italiano*, **lx**, 785–839; and Saccardo (ref. 8).

16. Chiarugi (ref. 15) and E. Chiovenda (1931), "Note sulla fondazione degli orti medici di Padova e di Pisa", *Atti dell' VIII° congresso internazionale di storia della medicina*, Pisa, 488–509, which cite the relevant earlier literature.

17. U. Viviani (1935), "La vita di Andrea Cesalpino", *Atti e memorie della R. Accademia Petrarca di lettere, arti, e. scienze*, N.S. **xviii–xix**, 17–84, 35.

18. J. R. Galluzzi (1781), *Istoria del Granducato di Toscana sotto il governo della casa Medici*, ed. seconda, Livorno, **iii**, 130–1.

19. *Petri Castelli...Hortus Messanensis*, Messina, 1640.

20. "Andò in questo tempo al mese di Maggio, havendo impetrato licenza dal Senato per causa pubblica, con molti scolari, vedendo prima tutte le valli da Padusa insino a Ravenna, nelle quali osservò la maggior parte delle piante degli ancichi e molte non da loro descritte. Si partì poi da Ravenna per Rimini, da Rimini all' Avernia, dove habitò S. Francesco, et ivi trovò molte piante non descritte dagli antichi. Ritornato a Rimini vide il giardino di Giulio Moderati, celebrato dal Matthioli..." From Aldrovandi's *Autobiography* written in 1586 printed in *Intorno alla vita...* (ref. 14), 1–27, 9–10.

21. G. Fantuzzi (1774), *Memorie della vita di Ulisse Aldrovandi medico e filosofo bolognese con alcune lettere scelte d'uomini eruditi a lui scritte, e coll'indice delle sue opere mss., che si conservano nella Biblioteca dell'Istituto*... Bologna, 166.

22. On *herbaria* in general see F. A. Stafleu (1967), *Taxonomic Literature*, Utrecht.

23. See esp. F. Rodriguez (1954–5), "Il museo aldrovandiano della Biblioteca Universitaria di Bologna", *L'Archiginnasio* **xlix 1**, 207–23. For the natural history paintings and drawings in Aldrovandi's collection see M. Bacci and A. Forlan (1961), *Mostra di disegni di Jacopo Ligozzi (1547–1626)*, Florence, and (U. Stefanutti), *Piante e animali nell'opera di Ulisse Aldrovandi* (Milan, s.d.). For further information on the emergence of natural history museums in Italy see the paper by A. G. Keller in the present volume.

24. C. Gesner (1972), *Historia plantarum*. Facsimileausgabe (eds. H. Zoller *et al.*), Zurich.

25. "Habeo metallica et fossilia plurima, et in dies ab amicis plura mittuntur; ex quibus, si intellexero quae adhuc tibi desint, ea libenter tecum sum communicaturus. Abs te vicissim rogo ut rarioribus seminibus et plantis meum studium adiuvare quoque velis, quae vero desidero hic seorsim notari..." Fantuzzi (ref. 21), 149–50.

26. "[In 1560] per seconda lettura lesse Theophrasto, *De causis plantarum*." *Intorno alla vita* (ref. 14), 11. Cf. Ms. Bologna, fondo Aldrovandi 78¹ described in Frati (ref. 14), 76. I have seen this ms. and it contains notes on Theophrastus.

27. Dallari (ref. 6) **ii**, 150 (for the year 1560–1).

28. "Emmi piaciuto de intendere dell'honorato grado, nel quale è stato eletto della lettione di filosofia naturale, delle piante, animali, e fossili, ordinariamente leggendo hor l'una, e hor l'altra intervallatamente, che certo sono lettioni di gran momento". Fantuzzi (ref. 21), 186–7 (dated 20 April 1561).

29. A. Baldacci, "Ulisse Aldrovandi e l'orto botanico di Bologna", *Intorno alla vita* (ref. 14), 161–72.

30. In addition to the *Hortus Messanensis* (ref. 16) see also his *Optimus medicus in quo conditiones perfectissimi medici exponuntur* (Messina, 1637). I plan to deal with these works in greater detail on another occasion.

31. This is generally well enough known to make it unnecessary to comment further. One byproduct of the anatomical revolution recently noted is that it may have led to a rapid decline in faculty psychology which was based on misconceptions concerning the anatomy of the brain. Though philosophers continued to endorse

the medieval scheme, medical men aware of the new anatomical discoveries knew better and abandoned it quite rapidly during the course of the sixteenth century. See K. S. Park (1974), "The Imagination in Renaissance Psychology" (M.Phil. Dissertation, University of London, Warburg Institute), 34–77.

32. M. T. Gnudi and J. P. Webster (1950), *The Life and Times of Gaspare Tagliacozzo*, *Surgeon of Bologna, 1545–99*, New York.

33. For a valuable recent survey see the article by T. B. Settle in *Dictionary of Scientific Biography* **ii**, 306–14.

34. H. B. Adelmann (1966), *Marcello Malpighi and the Evolution of Embryology*, Ithaca.

35. Further details and bibliography on this are given in my "Thomas Linacre and Italy" (ref. 5) and "Philosophy and Science in Sixteenth-Century Universities: Some Preliminary Comments", to appear in *Philosophy, Science and Theology in the Middle Ages* (eds. J. E. Murdoch and E. Sylla), Dordrecht-Boston, 1975.

36. See my paper cited in ref. 2.

37. A. Favaro (1922), "I lettori di matematiche nella Università di Padova dal principio del secolo XIV alla fine del XVI", *Memorie e documenti per la storia della Università di Padova* **i**, 1–70.

38. The term, along with *cosmografico* and *cosmografo*, was used fairly commonly in Italy from the fourteenth to the seventeenth century. See *Grande dizionario della lingua italiana* (Turin, 1961f.) **iii**, 887.

39. See Schmitt (ref. 2), 258.

40. A. Franceschini (1970), *Nuovi documenti relativi ai docenti dello studio di Ferrara nel secolo XVI*, Ferrara, 263.

41. P. L. Rose and S. Drake (1971), "The Pseudo-Aristotelian *Questions in Mechanics* in Renaissance Culture", *Studies in the Renaissance* **xviii**, 65–104, at 92–6.

42. The evidence for Moleto having taught the *Optica* in 1583–4 is given by A. Favaro (1917–18), "Amici e corrispondenti di Galileo Galilei: XL. Giuseppe Moletti", *Atti del R. Istituto Veneto di scienze, lettere ed arti* **lxxvii**, 47–118, at 63. The *prospectiva* listed for 1586–7 may also mean "optics" in this context. Moleto also left behind at least two mss. on optics [Milano, Biblioteca Ambrosiana, R.94.inf. and S.100.inf., not seen] as listed by Favaro, 87.

43. Cited by Favaro (ref. 42), 63 based on Riccoboni's *Orationes* (Padua, 1591), a volume which I have been unable to see.

44. The latter hypothesis is Favaro's (ref. 42), 63–4.

45. Evidence is in Schmitt (ref. 2), 260.

46. The recent discussion was initiated by Randall, who took the work of Ragnisco and Cassirer as his starting point. For a discussion of the subject with bibliographical references see C. B. Schmitt (1971), *A Critical Survey and Bibliography of Studies on Renaissance Aristotelianism*, Padua, esp. 38–46. More recent work on the subject includes F. Bottin (1972), "La teoria del *regressus* in Giacomo Zabarella", in *Saggi e ricerche su Aristotele...*, Padua, 49–70 and A. Poppi (1972), *La dottrina della scienza in Giacomo Zabarella*, Padua, in addition to the work of Galluzzi cited in the next note. Just as I write this there has appeared a new study which deals with the interplay of Galileian methodology with the continuity of Aristotelian doctrine at Padua during the seventeenth century. See M. L. Soppelsa (1974), *Genesi del metodo galileiano e tramonto del l'aristotelismo nella scuola di Padova*, Padua.

47. See Rose and Drake (ref. 41); G. Crapulli (1969), *Mathesis universalis*, Rome; G. Cosentino (1970), "Le matematiche nella *Ratio studiorum* della Compagnia di Gesù", *Miscellanea storica ligure* **ii**, 169–213; G. Cosentino (1971), "L'Insegnamento delle matematiche nei collegi gesuitici nell'Italia settentrionale", *Physis* **xiii**, 205–17; P. Galluzzi, "Il 'platonismo' del tardo Cinquecento e la filosofia di

Galileo", in *Ricerche sulla cultura dell'Italia moderna* (ed. P. Zambelli) (Bari, 1973), 37–79; G. C. Giacobbe (1972), "Il *Commentarium de certitudine mathematicarum disciplinarum* di Alessandro Piccolomini", *Physis* **xiv**, 162–93; G. C. Giacobbe (1972), "Francesco Barozzi e la *Quaestio de certitudine mathematicarum*", *Physis* **xiv**, 357–74; G. C. Giacobbe (1973), "Alcune cinquecentine riguardanti il processo di rivalutazione epistemologica della matematica nell'ambito della rivoluzione scientifica rinascimentale", *La Berio* **xiii**, 7–44. I have not yet been able to see Giacobbe's *Le opere di Pietro Catena sui rapporti tra matematica e logica* (Pisa, 1974).

48. Some information on Proclus is to be found in the publications of Crapulli, Galluzzi and Giacobbe, "Francesco Barozzi..." cited in ref. 47. Barozzi taught part of Plato's *Republic* at Padua in 1560 in his mathematics course. See *Francisci Barocci* (1566),... *Commentarius in locum Platonis obscurissimum et hactenus a nemine recte expositum in principio Dialogi octavi de Rep. ubi sermo habetur de numero geometrico de quo proverbium est, quod numero Platonis nihil obscurius*, Bologna.

49. *Federici Delphini* (1559),... *De fluxu et refluxu aquae maris subtilis et erudita disputatio. Eiusdem de motu octavae sphaerae*, Venice, cf. Favaro (ref. 37), 62.

50. Ms. Milano, Ambrosiana S.100.inf. (not seen) as cited by Favaro (ref. 42), 87–8.

51. Ms. Firenze, Biblioteca nazionale, Conv. soppr. B.10.480. I am presently preparing a fuller study of this text for publication elsewhere.

52. Particularly important in this respect are Petrus Catena (1556), *Universa loca in logicam Aristotelis in mathematicas disciplinas hoc novum opus declarat*, Venice [seen in BM copy 519.e.32(3), which lacks a title page]; Petrus Cathena (1561) [sic]... *Super loca mathematica contenta in Topicis et Elenchis Aristotelis nunc et non antea in lucem aedita*, Venice; and Joseph Blancanus [Biancani] (1615), *Aristotelis loca mathematica ex universis ipsius operibus collecta et explicata...*, Bologna. On Catena see Giacobbe, "Alcune cinquecentine..." (ref. 47) and on Biancani see Galluzzi (ref. 47).

53. Francesco Barozzi (1560), *Opusculum in quo una oratio et duae quaestiones: altera de certitudine et altera de medietate mathematicarum continetur*, Padua, and Alessandro Piccolomini (1547), *In mechanicas quaestiones Aristotelis paraphrasis... Eiusdem commentarium de certitudine mathematicarum disciplinarum...*, Rome. Filippo Fantoni, *An demonstrationes mathematicae sint certissimae*, ms. cited in ref. 51. On Piccolomini and Barozzi see the writings of Galluzzi, Giacobbe and Crapulli cited in ref. 47. I am currently preparing a study on Fantoni. For the moment see my paper cited in ref. 2.

54. For a further discussion of these points see my paper cited in ref. 35.

55. For a general survey see P. L. Rose (1973), "Humanist Culture and Renaissance Mathematics: The Italian Libraries of the Quartrocento", *Studies in the Renaissance* **xx**, 46–105. For the specific case of Archimedes see M. Clagett (1964), *Archimedes in the Middle Ages* I, Madison, 12–14. Professor Clagett is preparing for publication further studies on the diffusion and influence of Archimedes in the Renaissance.

56. C. B. Schmitt (1971), "Theophrastus", in *Catalogus translationum et commentariorum* **ii**, Washington, 239–322, where further references will be found.

57. The relevant document is printed in J. L. Heiberg (1896), *Beiträge zur Geschichte Georg Vallas und seiner Bibliothek*, Leipzig, 19.

58. Franceschini (ref. 40), 250.

59. O. R. Bloch (1971), *La philosophie de Gassendi*, The Hague, which lists the earlier literature on the subject.

60. C. B. Schmitt, "L'Introduction de la philosophie platonicienne dans l'enseignement des universités à la Renaissance", to appear in *Actes du seizième colloque international* [Tours]: *Platon et Aristote à la Renaissance*.

61. F. Purnell (1972), "Jacopo Mazzoni and Galileo", *Physis* **xiv**, 273–94 and, for more general information and additional bibliography, the same author's "Jacopo Mazzoni and His Comparison of Plato and Aristotle" (Columbia University Dissertation, 1971).

62. A full study of this subject is lacking but see B. Brickman (1941), *An Introduction to Francesco Patrizi's Nova de universis philosophia*, New York, and P. O. Kristeller (1964), *Eight Philosophers of the Italian Renaissance*, Stanford, 110–26.

63. The point has recently been made by D. Hay (1973), *Italian Clergy and Italian Culture in the Fifteenth Century*, London; The Society for Renaissance Studies, Occasional Papers, no. 1, esp. p. 6.

64. See R. Villoslada (1954), *Storia del Collegio Romano*, Rome, and Cosentino (ref. 47).

65. This is based on Dallari (ref. 6).

66. A. Franceschini (ref. 40), 264.

67. A. Fabroni (1791–3), *Historia Academiae Pisanae*, Pisa, reprint Bologna, 1971 **ii**, 123–7. This is also corroborated by the information found in Pisa, Archivio di Stato, Univ. 177–80.

4. Science in the Early Royal Society*

M. B. HALL
(Imperial College of Science and Technology, London)

In the seventeenth century the Royal Society occupied a unique place in the European Republic of Letters. To the world at large it was in the enviable position of being the recipient of royal patronage (few outside England realised how little this meant) and yet independent. Its independence meant that it alone decided upon its membership, so that it could bestow patronage in its own right, and its judgement was known to be quite uninfluenced by affairs of state. It was well organised, far better so than the Académie Royale des Sciences in this period, with a succession of secretaries who took the exchange of information seriously, and were responsible for maintaining a network of correspondence unrivalled for its time. One of the first of these, Henry Oldenburg, founded the first purely scientific journal in 1665, the *Philosophical Transactions*, which although strictly a private venture of its editor, yet carried the imprimatur of the Society, and utilised its correspondence. Perhaps most important characteristic of all, the Royal Society established a point of view, one might almost say a philosophy, of scientific method; although by no means all scientists were convinced that it was the proper approach to an investigation of nature, all scientists envied and admired its results. It was a well-developed form of empiricism, unique to the early Royal Society, not by any means always perfectly consistent, but always successful, providing a model which others might follow.

The Royal Society's intention to foster science based upon observation and experiment, at home and abroad, is clearly revealed in Oldenburg's correspondence. Thus early in 1663, when the Society made its first conscious endeavour to bring within its orbit foreign scientists of distinction, Oldenburg wrote to the Danzig astronomer, Johannes Hevelius[1]

> "it is now our business, having already established under royal favour this form of assembly of philosophers who cultivate the world of arts and sciences by means of observation and experiment, . . . to attract to the same purposes men from all parts of the world who are famous for their learning, and to exhort those already engaged upon them to unwearied efforts".

* An earlier version of this paper entitled "Empiricism and Rationalism in the early Royal Society" was read as a Clark Lecture at Scripps College, Claremont, California, on 11 February 1971.

To this Hevelius replied ten months later[2]

> "The learned world . . . owes humble and eternal thanks to the King of England, a prince worthy of every high praise, because he has founded a unique assembly of those philosophers who cultivate and advance the arts and sciences by following not tradition, but observations and experiments alone. For thus will hidden secrets be revealed at last and new miracles appear that were formerly concealed in the majesty of nature."

This empiricist approach was clearly aimed at from the very beginning. True, Wallis may have been a little influenced by later developments when he remembered that after 1645 the then youthful future Fellows were "inquisitive into natural philosophy and particularly of what hath been called the New Philosophy, or Experimental Philosophy",[3] but hardly that these first meetings involved those attending in "a weekly contribution for the charge of experiments".[4] This interest in experiment received decided emphasis in the First Charter where the King was made to declare[5]

> "We particularly favour those philosophical studies which endeavour by solid experiments either to extend new philosophy or to improve the old."

Indeed, the Society as a whole early felt that the Charters actually set limits to the Society's activities, limits that could be conveniently invoked when strangers or foreigners sent in letters and books which lay outside its prescribed scope. Thus in April 1663 Dr Eccard Leichner sent a book of his on education, together with some remarks upon his philosophical and theological interests.[6] When the letter was read at a meeting in the summer, it was conscientiously referred to Wilkins, Wallis, Pell and Hooke for consideration and reply, as was to become normal practice. In the event the reply was drawn up by Oldenburg, although approved by the Society as a whole; in it Leichner was firmly told that[7]

> "The Royal Society says it is not its concern to have any knowledge of scholastic or theological matters, for it is its sole business to cultivate knowledge of nature and useful arts by means of observation and experiment, and to promote them for the safeguarding and convenience of human life. These are the bounds to which the Royal Charter limits this British assembly of philosophers, which they think it would be improper to transgress."

This was a plain enough rebuke firmly echoing the express purpose of the first as well as the recently granted second Charter. But it was not only while the Society was young that these principles were regarded as needing support; years later incautious outsiders still received courteous but firm rebuffs when venturing to submit pure hypotheses for the Society's consideration. Thus early in 1672 a German physician named Hannemann wrote to ask if he could receive the Society's opinion "on the matter of sanguification, and how it is performed".[8] When the letter was read on 8 February 1671/2, it had to compete with a letter from Wallis on meteorology, one by Tommaso Cornelio on tarantulas (an excellent piece of zoology), Newton's paper on light and colours, and Flamsteed's observations of Jupiter's satellites. Nevertheless the Society listened politely, and ordered[9] Oldenburg to return their thanks "for his respect of the Society", adding "that it is not their custom to be hasty in delivering their judgement in any philosophical matters; but that all things of that nature are committed by them to observations and experiments frequently and carefully made". Again, in the spring of 1673 an obscure Irish priest, John Sainte Croix (or Sanbrucius) presented his *Dialectic* to the Society, describing it as a "Tract of Rational, Naturall, Transnaturall, and Morall philosophie, according to the principles and genuine Spirit of the great Doctor John Scotus, our learned Country man, so highly famed abroad, tho in this our age, scarce known at home".[10] The Society this time was so little interested that Sainte Croix heard nothing until he wrote to Oldenburg two months later, when he received the courteous, but this time totally discouraging reply, that the Society had ordered its Secretary

> "to give you their kind thanks for yr respect in sending ym one of yr books, and to assure you of their good affections towards you. Sir, [Oldenburg added] you cannot but know the End of yt Royall Institution to be, to promote Natural knowledge by Experiments, and yt in order there unto, among their other endeavors, they invite all Ingenious men everywhere to study the Book of Nature, rather than ye writings of witty men."

Not unnaturally, no more was heard of Sainte Croix. Cleverer men than he might be discomfitted by the Society's sharp appraisal of their claims. Thus in February 1673/4 there was read a letter from Christoph Sand of Hamburg, theologian and translator of several early volumes of the *Philosophical Transactions* into Latin, in which he advocated the view that pearls were superfluous oyster eggs.[11] The Society this time ordered, "that the writer of this letter be thanked, and desired to let the Society know what ground he had for the truth of the matter of fact", whereupon Sand somewhat shamefacedly admitted that he had his information at second hand, though he believed that his informant was trustworthy.[12]

Even more brusque was the treatment received in 1684 by the German chemist Johann Kunckel who dedicated his *Chymischer Probier-Stein* (Berlin, 1684) to the Society; when it was presented to the Society, Boyle had a Latin extract made, and Dr Slare gave an account of it in English. What the Society at large thought of it does not appear, but Boyle told the then Secretary, Aston, "that the Society had not been used to judge in these cases: that they were now making experiments, and were not come so far as to frame systems; that he was glad to see a controversy managed with so many good experiments, and by so able men".[13]

With this faint praise Kunckel was dismissed. Yet even so he had received more attention than the unfortunate canon of Périgord, Armand de Gerard, who in 1673 wrote a long letter to Oldenburg about his rediscovery of some

c

alchemical works and symbolic pictures by Raymond Lull, stimulated thereto, curiously enough, by reading the Latin *Philosophical Transactions*.[14] Oldenburg did not read *his* letter to the Society, and the memorandum for his reply is brisk and practical[15]

"Rec. le 9 janv. 74. Resp. le 14 janv. et demandé l'histoire naturelle de Perigord."

Occasionally the Society had the satisfaction of finding its admonitions taken quickly to heart. In 1673 Sebastian Wirdig, Professor of Medicine at Rostock, sent his new book *Nova medicina spirituum* to the Royal Society. Oldenburg wrote that the book would be judged in due course and the judgement transmitted to the author. (The book was in fact given to Petty,[16] who seems never to have made a report.) While awaiting events Oldenburg added a perfectly safe word of advice, being certain of the Society's views after more than ten years as its trusted servant,[17]

> "Meanwhile, since this Royal Assembly was founded in order to establish a solid and fruitful philosophy based on observations and experiments properly performed, it particularly desires and solicits the philosophers of all nations to address themselves to this task with their utmost care and industry, so that [laying aside all idle shadowy notions] they may make a thorough examination of NATURE, her phenomena and their consequences; may reveal her by means of experiments performed with irreproachable faithfulness and tireless zeal; and may with due scrutiny entrust whatever thence they may observe to the permanent memorials of learning. . . . You, most learned Sir, should strive in the future to contribute the strength of your intellect and should not cease from urging your colleagues and other persons throughout Germany from working towards the same end."

To this earnest injunction Wirdig meekly declared that he had only submitted his book so that

> "[if it may be done] [the Society] should illuminate the *Medicina Spirituum*, that it should put the book to a fair test by its rich store of observations. . . . In time I shall, with God's help, put out another edition in order to substantiate each and every one of my problems with the positive observations from which they have emerged and by which they are faithfully and solidly approached and demonstrated."[18]

Again and again over the years Oldenburg was called upon to emphasise the point that the Society was founded to uphold empiricism against rationalism, as its charter had proclaimed. Perhaps, in view of what was to happen later, it would be well to recall here the clear statement by Hooke, apparently written in 1663, very possibly as a draft of the Society's statutes:[19]

> "The business and design of the Royal Society is—
> To improve the knowledge of natural things, and all useful Arts, Manufactures, Mechanick practices, Engynes and Inventions by Experiments—[not meddling with Divinity, Metaphysics, Moralls, Politicks, Grammer, Rhetorick, or Logick]. . . .
> In the meantime this Society will not own any hypothesis, system, or doctrine

of principles of naturall philosophy, proposed or mentioned by any philosopher ancient or modern . . . ; nor dogmatically define, nor fix axioms of scientificall things, but will question and canvass all opinions, adopting nor adhering to none, till by mature debate and clear arguments, chiefly such as are deduced from legitimate experiments, the truth of such experiments be demonstrated invincibly.''

And in Hooke's version, there were to be "no debates held at the weekly meetings of the Society, concerning any hypothesis or principal of philosophy . . . except by speciall appointment of the Society or allowance of the President''.

The statutes as finally enacted were, fortunately, far less fierce than this. They merely provided[20]

"The business of the Society in their ordinary Meetings shall be, to order, take account, consider, and discourse of philosophical experiments and observations; to read, hear, and discourse upon letters, reports, and other papers, containing philosophical matters; as also to view, and discourse upon, rarities of nature and art; and thereupon to consider, what may be deduced from them, or any of them; and how far they, or any of them, may be improved for use or discovery.''

It was fortunate that the statutes were less confining than Hooke might have made them, for its meetings would have been much impoverished had the Society avoided all discussion of theory. Witness the early meetings in 1661, when experiments, objects, instruments, histories of trades, observations and rarities were all, indifferently, "brought in", and discussion was restricted to matters of fact, as the minutes printed by Birch clearly show. Had the Society limited itself so, it would have been an assembly of virtuosi, not a scientific society; a body encouraging naive Baconism, not one promoting the sophisticated exploration of nature. Very few men of dogmatically empirical cast of mind ever were Fellows of the Royal Society; the nearest they got was having their letters read at a meeting, and receiving thanks and encouragement through the Secretary. Such a man was Joshua Childrey, rector of a quiet Dorset village, author of *Britannia Baconiana* (1660), a collection of random facts about various parts of England; he told Oldenburg in 1669 that

"Some 2 yeares before ye happy returne of the King, I bought me as many Paperbookes of about 16 sheets a piece, as my L. Verulam hath Histories at ye end of his *Novum Organon*, into which Bookes [being noted with ye figure & title given them by my Lord] I entred all Philosophicall matters, that I met with observable in my reading; & intend [God willing] to continue it. This I acquaint you with, to let you see how earnest & serious I have been for severall yeares in that that is ye businesse of ye R. Society.''[21]

This approach could and did produce useful matters of fact, as about tides, which could be seized upon by those like Wallis who interested themselves in a theory of tides (indeed Wallis had published his *Hypothesis about the Flux and Reflux of the Sea* in 1666),[22] but it was hardly what the Royal Society really expected of its Fellows. However much the Society as a body might hesitate to favour hypothesis, its aim was to establish something more than a mere collection of random experiment. Mere matter of fact was not valued for itself, but for the light it could shed on the Society's object, the establishment of a true philosophy of nature. So when in 1673/4 mathematician Sluse of Liège sent the Society a detailed answer to its many queries about the mineral spas of Belgium, Oldenburg wrote in his letter of thanks,[23]

> "We believe this to be the proper way for founding, some day, a true history of nature, by purifying author's accounts of whatever in questions of physics may have been the product of partisan zeal or reliance on the trustworthiness of others . . . the Society . . . deliberately draws the learned and skillfull of every country into its assembly, in order to pursue its purpose more successfully."

Some kinds of hypothesis had always been welcome—those "grounded in fact", and subject to confirmation by (empirical) facts which the Society was always ready to "entertain" and discuss. Individual Fellows might hold what views they liked, and did so—Cartesian, Baconian, Epicurean, mystic, Aristotelian, medical, chemical—the list might be nearly endless. What concerned the Society as an organised body was to avoid *its* commitment to *a priori* systems, those based upon empirically untestable tenets or principles. Such systems, they all held, prejudiced the minds of those who held fast to them, necessarily obscured their judgements, and led their proponents away from the only true philosophy, that of principles based upon empiricism. Now it would clearly not be true to say that upholders of such systems could not or did not practice experiment, for they manifestly did, and indeed often appealed to empirical evidence to support their beliefs. Aristotelians, Cartesians, Epicureans all might be, and in the later seventeenth century usually were, given to the performance of experiment almost as much as any dogmatic Baconian; indeed they often were Baconians, and vindicated Bacon's belief in the advantages of simple empiricism by making experimental discoveries. The crux of the matter came in the application of experiment, observation and discovery to the development of hypothesis and theory. The rationalist used his observations and discoveries to support a system or hypothesis already manifest (to him) on rational grounds; he must always find it easier to stretch his hypothesis or re-work his facts than to abandon it easily in the face of what he regarded as "mere" empirical evidence. The Royal Society strove to maintain an exalted empiricism, in which hypothesis must stand or fall by empirical fact. Hence it was bound to try at least to avoid speculative systems. In so doing wrangling over irreconcileable hypotheses was kept to a minimum, but this was only an incidental gain. At a higher level there was every hope that true, "solid" theory would benefit, and that all would be immediately aware of its advantages, and work to promote it.

At the same time individual Fellows in their private capacity might do as they liked, provided they did not profess to uphold their views as representing those of the Society. Newton was quite correct when he noted dryly "I have observed the heads of some great virtuosos to run much upon hypotheses".[24] Over the years they had indeed done so, both publicly and privately. As early as 4 September 1661 Kenelm Digby had read a letter from Frenicle de Bessy, setting forth his "hypothesis of the motion of Saturn", which was received with interest.[25] He was thanked by the Society and far from being rebuked for his hypothesis; on the contrary, Sir Paul Neile quickly (and rightly) pointed out that Frenicle's hypothesis was very similar to that developed by Christopher Wren some years earlier. Wren was present at the meeting, and when pressed, agreed to draw up a letter giving details of the ideas which he had developed on the subject (many in a lecture at Gresham College, where he was still Professor of Astronomy), and also to produce a short Latin hypothesis. He was not much interested for his own part; he was not a man to concern himself with priority disputes, and in any case had already abandoned his hypothesis upon reading Huygens' *Systema Saturnium* in 1659, recognising immediately the superiority of Huygens' views. Sir Robert Moray, Huygens' usual correspondent at this period, wrote to Huygens for his opinion, and his reply was read (together with (Wren's) at the meeting of 9 October.[26] Here the matter rested until, ten years later, Saturn's ring emerged from its edge-on, virtually invisible position, when astronomers once again took up the problem, and Huygens' hypothesis was, as he proudly proclaimed, successfully vindicated by observation[27] and thereby shown to be worthy of a Fellow of the Society, as Huygens was. This was an example of what all the Society recognised as a successful hypothesis—or as one would now say, theory.

It was indeed a customary and regular procedure of the Royal Society to take well-founded hypotheses seriously, to the extent of referring them to one or more Fellows to "consider on" and report upon. After all, the Society existed to promote natural philosophy everywhere, and although it was easiest to encourage purely empirical science, because it was quickly possible to evaluate its worth, the Society by no means limited itself to praising those who submitted observation and experiment. True, it welcomed such work from, for example, Malpighi, urging him to further endeavour and lending him assistance in publication, after having reflected upon the value of the work he submitted. The same reception might be accorded to theoretical work. Consider the case of Sluse, already a distinguished mathematician, well known to a number of Fellows including Wallis. Oldenburg opened a correspondence with him[28] in February 1666/7, beginning "I think it proper, famous Sir, for a man who loves learning and is familiar with the laws governing the learned world to address those who cultivate the arts and promote the sciences, wherever they live". He went on to explain the design of the Royal Society in his usual terms: "it aims at building up a solid system of philosophy fit to explain the true causes of natural phenomena and really promote the amenities of human life, on the basis of observations and experiments frequently and accurately performed", and to urge him to contribute. Sluse responded politely but uninformatively, and Oldenburg immediately replied by sending him a mathematical paper for comment.[29] This was the commencement of a long and constructive correspondence, wherein Oldenburg acted at times as a transmitter for John Collins, at times for Wallis, sometimes for Huygens and sometimes for Sluse himself, promoting fruitful mathematical interchanges. The Society's approval is amply indicated by its electing Sluse a Fellow in 1674. Although occasionally questions in chemistry, mining, metallurgy and natural history were sent to him (to all of which he replied) most letters were mainly concerned with pure mathematics. Indeed, Sluse was the first Continental mathematician to learn (in 1669) of the highly original mathematical work of the young Isaac Newton,[30] and Sluse and Newton were mutually polite about each other's prowess in developing methods of finding tangents to complex curves. English mathematics was, at this period, highly developed, and mathematicians everywhere received very considerable encouragement from the Royal Society.

This was notably true of the young Leibniz. He first appears on the English scene in the summer of 1670 as an unknown young jurist and councillor of the Elector of Mainz with an obscure mathematical work (*De arte combinatoria*, 1666) to his credit, and with an interest in the theory of motion, general problems in natural philosophy and in the problem of a universal language.[31] Under encouragement from Oldenburg, Leibniz had his hypothesis printed under the general title *Hypothesis physica nova* (Mainz, 1671) and Part I, dedicated to the Royal Society under the title *Theoria motus concreti*, was sent to England for approval in the following March.[32] Leibniz could send only one copy, so it was reprinted in England and at the meeting of 23 March 1670/1 "It was ordered, that Mr. Boyle, Dr. Wallis, Dr. Wren, and Mr. Hooke should be desired to peruse and consider this book, and report their sense of it to the society, in order that a proper answer might be returned to the author". Wallis wrote a full Latin letter to Oldenburg; when this was read to the Society

> "It was ordered, that the doctor should receive the thanks of the society for this good account; that his letter should be entered, and hereafter compared with the sentiments of those other persons, who had been desired to consider the *hypothesis*; and that thereupon an answer should be written by the secretary to Monsr. Leibnitz, containing the judgment of the said persons concerning that *hypothesis*".[33]

Wallis presented a very fair critique; he was, as he acknowledged, favourably disposed towards an hypothesis which was not very dissimilar from his own, or, as he put it, "As for the work itself, I find many things expressed in it with very good reason, and to which I can fully assent since my own views are the same". These he proceeded to enumerate, adding, "There are many other things it is needless to recite here which in my opinion are said with great probability, if not with certainty". But having given praise to particular arguments, he went on

"I shall say nothing immediately about the hypothesis as a whole, for the reason at least that I am not apt to give my assent to a thing newly proposed (at any rate in physics, if not mathematics), either until it appears more directly through the arguments of the learned on both sides of the question what is to be thought of it, or until the truth emerges through the very clearness of the thing, as happens not rarely with true hypotheses. . . . The question [the cause of cohesion] must be left to time and the arguments of the learned on either side. Indeed, almost the same things happen with new hypotheses as with the swings of pendulums; after many oscillations either side, at last they come to rest perpendicular. We have seen this with the Copernican hypothesis. . . . The same may be said of Harvey's circulation of the blood . . . [and, of many other discoveries in physics and even anatomy] And the same is to be expected in this affair, and in the case of other new hypotheses that can be proved neither by ocular inspection nor by certain demonstration, so that if they are founded on true reasoning they will at last (but not without wrangles on both sides) find a place in the minds of those who philosophise freely; meanwhile they remain in suspense."

This excellent sketch of the English view of hypotheses was published in the *Philosophical Transactions* in its original Latin, and so was available for all to read. Hooke's opinion of Leibniz's work was less judicious: he reported "that he was not satisfied with it",[34] and of the second part (*Theoria motus abstracti*) he spoke even more severely, telling the Society that "he had perused and considered [it], but was of opinion, that [Leibniz] had not hit right".[35] Oldenburg was thereupon instructed to send it to Wallis, whose reply was again to be published in *Phil. Trans.*, but this time Wallis refused to comment at length, partly because he said he was very busy, partly because "it may seem invidious to criticise the writings of others", so he only paused to remark that he was not satisfied with Leibniz's explanation of cohesion, which he himself had discussed years ago with William Neile. He preferred, he said, to "keep silent and pronounce no opinion prejudicial to others, so that everyone may freely adopt whatever view shall seem to his judgement most agreeable to reason".[36] Oldenburg reported to Leibniz what had been done in respect of his work, informing him that

"the Society [as its custom is] commended both books, on different occasions, to certain mathematicians and natural philosophers to be read and considered. What was done on this occasion was what is usually done when judgements have to be reported about matters lying outside the rigour of mathematics; for indeed those philosophers diverge into a variety of opinions."

Oldenburg then quoted from Wallis's two letters, and informed Leibniz that he was having the two little tracts reprinted so that the Society "might seek out the opinions of our men of learning far and wide and perhaps . . . borrow some light from them where you so far have seen but darkly".[37]

In pursuance of this aim, Oldenburg sent a copy of the London edition to the Jesuit Pardies, along with a copy of Oldenburg's English translation of Pardies' *Discours du mouvement local*, remarking

"Since this author gives us his ideas about both abstract and concrete motion (the subject which at present occupies the philosophers of Europe), we believe it is necessary to examine everything written on this subject, so important for all philosophy; it was this, Sir, that made me judge that you would be very glad to compare this work with yours, and with those of Messieurs Wallis, Wren and Huygens. When you have done this you will very much oblige me if you will tell me your opinion of it."[38]

The reference here is to the long discussion of "the laws of motion" under the Society's aegis in 1668; this had begun with Hooke's proposing "that the experiments of motion might be prosecuted, thereby to state at last the nature and laws of motion" and had led to the extraction of papers from Wallis, Wren and Huygens, all ultimately printed in the *Philosophical Transactions*.[39] Pardies' reply to Oldenburg's request suggests that either he had more faith in the views of established scientists than of those of novices, or else a deep belief in the virtues of English science, for he wrote[40]

"I have read the little book *De motu abstracto et concreto*. You will excuse me, please, from telling you my opinion of it. I should not make the same difficulty with respect to the book by Mr. Wallis or that by Mr. Wren if I had read them, for I am already strongly convinced of the excellence of their intelligence and of their profound learning."

Charitably, Oldenburg told Leibniz neither of this rebuff, nor of that by Hooke, but rather encouraged him to turn to mathematics, especially after his exhibition of his calculating machine to the Society on his London visit in 1672/3, and his development of mathematical interest during his sojourn in Paris thereafter.[41]

As in physics, so in astronomy, the Royal Society was ready to consider well-founded hypotheses. These Oldenburg produced for discussion by the Society, whereas those he judged less well of (like Cocherel's schemes for determining longitude astronomically) he dealt with by personal consultation.[42] At the Society's meeting of 26 June 1672 Oldenburg read a letter from Cassini[43] which was "accompanied with a written paper of two sheets in folio, giving an account of his endeavour for settling an hypothesis of the motion of Jupiter and his satellites". The minutes continue, "This paper was committed to the perusal and consideration of Mr. Hooke, who was desired to make a report of it to the Society at their next meeting, especially as the author expressed his desire of having the sense of the Society, or some members thereof, upon the said paper".

After some prodding, Hooke on 10 July gave the desired report which characterised Cassini's "system" as "considerable, and deserved to have good notice taken of it in the observations of the motions of those stars". Most unfortunately neither Cassini's paper (although later sent to other astronomers) nor Hooke's account nor Oldenburg's letter to Cassini now survives.[44]

Similarly the Society eagerly promoted discussion of Hooke's interesting proposal of measuring stellar parallax, published in 1674 as *An Attempt to Prove the Motion of the Earth from Observations*. At the end of March 1674 Oldenburg sent copies to both Cassini and Hevelius explaining[45] that these observations "seem to show that there is a sensible parallax of the earth's orbit to a fixed star in Caput Draconis", adding carefully "not that the author makes it appear that the matter has been entirely settled by him, but that he believes himself to have embarked on the right method by which that noble controversy may be settled at last". Huygens very quickly responded favourably; his remarks as published in the *Philosophical Transactions* were all that anyone could wish, giving both praise and the offer of assistance from the Paris Observatory.[46] He added, "This, if it succeeds, will prove an almost entire conviction of the *Anti-Copernicans*, since there will remain for them nothing but this un-grounded subterfuge, to say, that the Center of the Sphere of the Fix't Stars continually changes its place for an Annual Motion".

Similarly Cassini praised both Hooke's method and his choice of star.[47] He noted cautiously, "Our observations have not, indeed, so far disclosed anything so obvious of this sort", but promised to assist in further observations. The most cautious reaction was that of Wallis, who wrote[48]

> "Mr Hooks observation . . . I am very pleased with. Onely I am sorry it hath been no further prosecuted. For depending as it were upon little more than one observation and that a nice one: they who like not the thing, will rather content them selves to attribute it to some errour, from some undeeded accident, than take ye pains to repeat it."

Wallis's caution was indeed wise, both in respect of method and in respect of the observation, for it was not repeatable. But the Society's position was irreproachable, for further observations alone could determine the value of Hooke's observation, and the method itself was perfectly sound in principle, if not in practice.

Many other examples of the discussion of hypotheses could be given, in which Hooke often played a part, for he was excellent in discussion, and quick to develop ideas on first hearing them. Thus on 12 November 1674 Wallis read his "discourse of gravity and gravitation grounded on experimental observations", and Hooke, among others, proposed some discussion of "Springiness" and its cause, which he promised to report upon, and to show how an elastic body could be composed of non-elastic particles. He was encouraged to do so.[49] Again on 4 February 1674/5 Edmund King presented a theoretical paper on the structure and motion of muscles, and Hooke, together with Grew and Croune, was lively in discussing muscle structure. Against this background it is easy to see why Newton, having "observed the heads of some great virtuoso's to run much upon Hypotheses", offered his "Hypothesis explaining the Properties of Light discoursed of in my severall Papers" at this time.[50] Much of the difficulty he had earlier experienced in 1671/2 came, at least in his view, from the predilection of his critics to lean

C*

too heavily upon hypotheses, to fail to distinguish hypothesis from proved truth, and to demand hypothesis where none was intended. And in this there was not much to choose between English and Continental scientists.

The story is well known, but is well worth reconsidering for the light it throws upon the English attitude towards scientific method. When on 6 February 1671/2 Newton, at the Society's request, wrote to Oldenburg about the surprising results of his experiments with prisms, he did so in a carefully empirical style.[51] Most of the letter is a simple description of experiments performed, and so cautious was Newton that he even refers to the sine law of refraction as an hypothesis although, he admitted, "by my own & others Experience I could not imagine it to be . . . erreneous". There follows the recital of various "suspitions" about the behaviour of light rays whose "gradual removal . . . at length led me to the *Experimentum crucis*", which he describes briefly and without a diagram (so that many found it difficult to comprehend). Whence he triumphantly concluded "And so the true cause of the length of that Image was detected to be no other, then that *Light* consists of *Rays differently refrangible*". Only after considering the effects of this discovery upon "Glass-works" (the making of telescopes and microscopes) does Newton, again cautiously, approach "*the Origin of Colours*". Carefully he noted,

> "A naturalist would scearce expect to see ye science of those become mathematicall, & yet I dare affirm that there is as much certainy in it as in any other part of Opticks. For what I shall tell concerning them is not an Hypothesis but most rigid consequence, not conjectured by barely inferring 'tis thus because not otherwise or because it satisfies all phaenomena (the Philosophers universall Topick,) but evinced by ye mediation of experiments concluding directly & wthout any suspicion of doubt . . . I shall . . . lay down the *Doctrine* first, and then, for its examination, give you an instance or two of the *Experiments*, as a specimen of the rest."

It must be remembered that Newton was at this time still a young man, who had been Lucasian Professor for only two years, recently elected to the Royal Society (on 11 January 1671/2) and little known even among its Fellows. He might well take seriously the *Nullius in verba* of its motto, and its professed intention of avoiding all *a priori* systems and hypotheses. His use of the word "doctrine" is not an empty avoidance of loaded words like hypothesis and system; the word (a favourite with him about this time) indicated a degree of certainty taken to result from thorough experimental examination. The "doctrine" here was a careful series of empirical statements which led him ineluctably to the view that white light is composed of rays of different colours, and that the mixing of these rays gives rise to other colours, all together producing white.

The letter was read at the Society's meeting of 8 February, where it was evidently received with acclaim. The minutes report that

> "It was ordered, that the author be solemnly thanked, in the name of the Society, for this very ingenious discourse, and be made acquainted, that the Society think

very much of it [and proposed] if he consent to have it forthwith published, as well for the greater convenience of having it well considered by philosophers, as for securing the considerable notions of the author against the pretensions of others".

Oldenburg's report shows that the formal minutes do not give more than a hint of the warmth of the paper's reception. He wrote[52] "I can assure you, Sir, that it . . . mett both with a singular attention, and an uncommon applause". As usual, it was "ordered" that several Fellows "be desired to peruse and consider it, and bring in a report to the Society"; these were named as Seth Ward, Boyle and Hooke. The latter hastened to bring in a report at the very next meeting on 15 February, a report so filled with harsh and ungenerous criticism as to embarrass the Society. As the minutes tell us, it was decided to send Hooke's paper to Newton but not to print it at the same time as Newton's own paper "lest Mr. Newton should look upon it as a disrespect, in printing so sudden a refutation of a discourse of his, which had met with so much applause at the Society but a few days before". Its printing was to be delayed until Newton had time to reply. Unfortunately Oldenburg's letter reporting this, and conveying Hooke's paper has disappeared, so we cannot tell in what words he expressed the dismay felt by the Fellows when their respected and mature Curator ungraciously attacked a young and promising newcomer.

Hooke's criticism was not only harsh and rude, but displays a tenacious adherence to hypothesis which seems surprising in a professed empiricist. But empiricism wears many faces, and although Hooke could dismiss others' *a priori* hypotheses easily enough, like most people he found it difficult to dismiss his own. He began with apparent politeness but with an insistence upon hypothesis which Newton was to find tendentious and troublesome;[53]

> "I have perused the excellent discourse of Mr. Newton about colours and refractions, and I was not a little pleased with the niceness and curiosity of his observations. But, tho' I wholly agree with him as to the truth of those he hath alledged, as having, by many hundreds of trials, found them so, yet as to his hypothesis of salving the phaenomena of colours thereby, I confess, I cannot see yet any undeniable argument to convince me of the certainty thereof."

A resounding snub, this, denying the originality of Newton's observations as well as taking his doctrine, so carefully developed upon their basis, as mere, and unproved, hypothesis. Having firmly declared that Newton's "hypothesis" was *not* supported by Newton's experiments, Hooke then proceeded to declare that these experiments in fact confirmed what his own had supported—his own hypothesis.

> "For all the experiments and observations I have hitherto made, nay and even those very experiments which he alledged, do seem to me to prove that light is nothing but a pulse or motion propagated through an homogeneous, uniform and transparent medium: And that colour is nothing but the disturbance of that light

by the communication of that pulse to other transparent mediums, that is by the refraction thereof. . . . But how certain soever I think myself of my hypothesis, which I did not take up without first trying some hundreds of experiments; yet I should be very glad to meet with one experimentum crucis from Mr. Newton, that should divorce me from it. But it is not that, which he so calls, will do the turn; for the same phaenomena will be salved by my hypothesis as well as by his without any manner of difficulty or straining: nay I will undertake to show an other hypothesis differing from both his and mine, that shall do the same thing."

Nothing could more plainly show that Hooke did not understand Newton's approach at all than his conclusion, "If Mr. Newton hath any argument, that he supposeth an absolute Demonstration of his theory, I should be very glad to be convinced by it". For what could be an absolute demonstration, when Hooke was prepared to argue that he could think up yet other hypotheses which should explain Newton's observations, besides that one which he already held? How *could* there be an experimentum crucis, when (it is obvious) Hooke's criteria for the acceptance of an hypothesis were rational, *a priori* ones? No doubt he could think of other hypotheses, but that was hardly the point. Hooke's hypothesis was the sort of which the Royal Society, and he himself indeed, had long disapproved. Clearly, he could not see that Newton's "doctrine" was precisely the sort of physical theory, arising directly from empirical evidence, which they professed to welcome. Nor that the mere performance of *more* experiments ("some hundreds") was unlikely to alter the position.

Newton's first reply was simply to say[54]

"I received your[s] . . . And having considered Mr Hooks observations on my discours, am glad that so acute an objecter hath said nothing that can enervate any part of it. For I am still of the same judgment & doubt not but that upon severer examinations it will bee found as certain a truth as I have asserted it."

He was understandably reluctant to answer Hooke at length, since their fundamental disagreement turned on questions of scientific method, and others, like Huygens and Pardies, had raised more cogent criticisms of fact and presentation. And probably he was correct; for Pardies was soon convinced, and while Huygens, like Hooke, never ceased to regard Newton's theory of colours as unproven, he went much closer than Hooke in accepting Newton's results.

Strongly pressed by Oldenburg acting at the behest of the Society, Newton finally sent up a reasoned reply for eventual publication in the *Philosophical Transactions*,[55] even though Hooke's arguments seemed to him to "consist in ascribing an Hypothesis to me wch is not mine; in asserting an Hypothesis wch as to ye principall parts of it is not against me; in granting the greatest part of my discourse is explicated by that Hypothesis; & in denying some things the truth of wch would have appeared by an experimentall examination". Newton examined Hooke's arguments carefully, trying to point out the difference between Hooke's hypotheses and his own doctrine. Of the experimentum crucis he could only say that "it was such", adding "I cannot be convinced of its insufficiency by a bare denyall without assigning a reason for it". Clearly he despaired of convincing Hooke of his own conviction that "I do not think it needful to explicate my Doctrine by any Hypothesis at all". Well might he say "I must confesse at ye first receipt of those Considerations I was a little troubled to find a person so much concerned for an Hypothesis, from whome in particular I most expected an unconcerned & indifferent examination of what I propounded".

No doubt that is what Hooke had intended to provide; it is certainly what he would have advocated; the upshot firmly proved what the Society had long held, that addiction to *a priori* hypotheses both provoked dissension and prevented the uninterrupted advancement of truth. That Hooke could not perceive his own non-empirical bias is clearly demonstrated in his draft reply, never read to the Society nor seen by Newton, wherein he proclaimed[56] "I judge noething conduces soe much to the advancement of Philosophy as the examining of hypotheses by experiments & the inquiry into Experiments by hypotheses. and I have the Authority of the Incomparable Verulam to warrant me."

Hooke's Baconianism was fervent, but Newton's position was more soundly empirical, for in his mind experiments offered proof, and once they had proved an hypothesis, it was not to be set aside in favour of another just for the intellectual exercise. Hence the Fourth Rule of Reasoning in Natural Philosophy of the *Principia*—Newton by the time he wrote it knew all too well how easily the "argument from induction [might] be evaded by hypotheses" otherwise, and how difficult it was for many of his contemporaries to resist devising hypotheses when none was required. Or as he put it in 1675 "I have . . . found, that some when I could not make them take my meaning, when I spake of the nature of light and colours abstractedly, have readily apprehended it when I illustrated my Discourse by an Hypothesis".[57]

Hence (especially when urged by many of the Society) he produced his *Hypothesis of Light* with, however, the very necessary caution

> "I shall not assume either this or any other Hypothesis, not thinking it necessary to concerne my selfe whether the properties of Light, discovered by me, be explained by this or Mr Hook's or any other Hypothesis capable of explaining them; yet while I am describing this, I shall sometimes to avoyde Circumlocution & so represent it more conveniently speak of it as if I assumed it & propounded it to be believed".

This was plain enough—and to reinforce it in his covering letter (also read to the Society) he carefully stated[58] "Sr. I had formerly proposed never to write any Hypothesis of light and colours, fearing it might be a means to ingage me in vain disputes: but I hope a declar'd resolution to answer nothing that looks like a controversy (unles possibly at my own time upon some other occasion) may defend me from yt fear".

Perhaps curiously, this occasion when Newton professed an hypothesis went off more smoothly than the previous occasion over three years before, when he had not. The Society declared itself "well pleased". Hooke went so far as to write a placatory letter to Newton, disclaiming the desire to quarrel in print or public, and insisting that he valued Newton's work exceedingly.[59] Although the letter was written a week after the reading of part of Newton's "Hypothesis", Hooke rather curiously had not yet read it himself, nor, apparently, absorbed it, and reserved the right to send any objections privately to Newton. Newton generously agreed with Hooke, saying,[60] "There is nothing wch I desire to avoyde in matters of Philosophy more then contention, nor any kind of contention more then one in print" and warmly praising Hooke's work. Yet Hooke at the very next meeting of the Society[61] proposed that *his* hypothesis would explain Newton's observations and experiments as well as Newton's hypothesis could do, and Newton's patience was soon exhausted. Not unnaturally, he refused to allow his last paper to be published, fearing acrimony. It came hard on him to be censured by Hooke for following the rules of the Society of which they were both Fellows more strictly than Hooke, the senior, was prepared to do. Clearly they were both Baconians; equally clearly they interpreted the Society's emphasis on empiricism in different ways. To the outside world, however, the difference was little apparent at the time, and so in the 1670s, as later, Newton's work was regarded abroad as quintessentially English and empirical.

Only one man rivalled him in this respect: Robert Boyle, as much older than Hooke as Hooke was than Newton, already an established scientist with an international reputation when Newton was still an undergraduate. There is indeed some reason to believe that Boyle, who had been endeavouring to build up a body of evidence to support the mechanical philosophy since 1654, at least, was finally led to publish *New Experiments Physico-Mechanicall, Touching the Spring of the Air, and it Effects* (Oxford, 1660), as a witness to the world of what the new English science was accomplishing. As he remarked in his preface to the reader[62]

> "intelligent persons in matters of this kind persuade me, that the publication of what I had observed touching the nature of the air, would not be useless to the world; and that in an age so taken with novelties as is ours, these new experiments would be grateful to the lovers of free and real learning: so that I might at once comply with my grand design of promoting experimental and useful philosophy, and obtain the great satisfaction of giving some to ingenious men".

So anxious was Boyle on this occasion to emphasise the advantages of *experimental* philosophy, and so eager was he to demonstrate its advantages, and its niceties, that he carefully avoided any prolonged discussion of the nature of air, such as might have been expected considering the nature of the treatise. But underlying the whole intensely empirical work is the assumption of the truth of the broad principles of the mechanical philosophy. This view and method is more clearly expressed in his next work, *Certain Physiological Essays*, which is composed half of short treatises on the methodology of experiment, and half of treatises supporting the mechanical philosophy by means of specific experiment. And here, in the prefatory remarks to what he regarded as the most convincing empirical proof of the existence of corpuscles of matter (*The Essay on Nitre*) Boyle carefully explained his policy of utilising in discussion a neutral view of the mechanical philosophy which might be acceptable to Cartesians, Epicureans—and himself.[63] What mattered most to him at this stage was to show how chemical experiments could "illustrate" as he said, the corpuscular philosophy—in later life he successively would claim they could "demonstrate" and even "prove" it. This was to him, as to the Royal Society at large, the true intention of the Society's purpose and plan.

At one level this was perfectly successful. Boyle's experimental technique was superb, his experimental imagination always fertile; his books were eagerly awaited at home and abroad, Latin translations sold even better than the original English editions, and Boyle's reputation as an experimental scientist was unrivalled in the Republic of Letters.[64] Scientists and scholars from Paris to Hamburg were constantly on the watch for his books, which were seized upon as they came in the booksellers' shops, and promptly reviewed. Few would have disagreed with Christiaan Huygens' father Constantijn, who in the spring of 1674 wrote to Oldenburg[65]

"L'eau me vient fort à la bouche des belles choses que vous dites que Monsieur Boyle nous promet. Comment est-il possible que tant de solide se niche dans un seul esprit? je vous prie de me bien conserver toutes les bonnes graces d'un si eminent personnage, qui merite mieux ce titre que par ensemble tout gl'Eminentissimi di Roma, che mai furono e mai saranno."

It was for his experiments that Boyle was principally valued, but his life-long support for the mechanical philosophy was also highly esteemed. It is therefore of some interest to examine what two of the most eminent rational philosophers of the age—Spinoza and Leibniz—thought of Boyle's empiricist support of the mechanical philosophy which they themselves accepted and supported on rational (Spinoza would have said mathematical, meaning rigorous) grounds.

It should be said at the outset that there was no difference as to the validity of the mechanical philosophy, broadly interpreted, between Boyle and Spinoza or Leibniz. On the contrary, the difference arose solely from a point of method or, put another way, about the fundamental nature of hypotheses: are they demonstrable logically and therefore "rigorously" from a set of *a priori*, intuitively certain principles, or should and must they be demonstrated by empirical evidence.[66] When Spinoza read Boyle's *Certain Physiological Essays* in its Latin edition, and particularly the *Essay on Nitre*, he was struck by two points, the most important being that Boyle tried to prove by experiment such things as the particulate structure of matter. As he said[67]

"Mr. Boyle tries to show that all tactile qualities depend only upon motion, shape, and the remaining mechanical states. Since Mr. Boyle does not put forward his proofs as mathematical, there will be no need to enquire whether they are altogether convincing. . . . One will never be able to prove this by chemical or other experiments, but only by reason and calculation. For by reason and calculation we divide bodies infinitely, and consequently the forces which are required to move them; but we shall never be able to prove this by experiment."

Spinoza's view was that Boyle's empiricism was a waste of time: the particulate structure of matter and the mechanical philosophy had already been quite amply demonstrated logically by, as he says, Bacon and Descartes, whose arguments were perfectly satisfactory, so why try to educe these arguments from experiments? And if Boyle merely wished to illustrate the mechanical philosophy from experiments, why bother to make new ones, when there were plenty of well-known experiments to hand? The other point that specially struck Spinoza was that Boyle seemed to differentiate between chemical cohesion and physical cohesion, between a chemical compound and a physical mixture and he could not believe that this difference was valid. Indeed no rational argument could penetrate the fundamentals of empirical chemistry, and so Spinoza was led to doubt that Boyle could be arguing that chemical combination was something different from mixture, or that Boyle's chemical insights were valid. Through Oldenburg, Boyle patiently reasserted the importance and validity of what he was then trying to do—what he would continue to try to do for the next thirty years—and tried to show that Spinoza's hypotheses were gratuitous and unproved; but he had little hope of convincing Spinoza to abandon reasoning *more geometrico*, and wisely soon ceased to try.

Spinoza's ideas were to receive a new airing through the publication of much of his correspondence in his *Opera posthuma* of 1677. It was from thence that they came to the attention of Leibniz, who accepted Spinoza's views on hypotheses both as reinforcing those of Descartes and as convincing in themselves. In commenting upon Locke's *Essay on Human Understanding* Leibniz was inevitably led to consider what he called "L'Art de decouvrir les causes des phenomenes, ou les hypotheses veritables"; he remarked that Boyle was an excellent writer on experimental method, but thought that what was required was some sure means of drawing consequences from experiments—without which one would only learn what "un homme d'une grande penetration pouvoit decouvrir d'abord", as both Descartes and Spinoza had noticed. Of Boyle Leibniz remarked sadly[68]

[Il] s'arreste un peu trop, pour dire la verité, á ne tirer d'une infinité de belles experiences d'autre conclusion, que celle qu'il pouvoit prendre pour principe, savoir que tout se fait mecaniquement dans la nature, principle, qu'on peut rendre certain par la seule raison, et jamais par les experiences, quelque nombre qu'on en fasse."

And this was the conclusion that rationalists were bound to draw. Nevertheless, they admired Boyle for the "infinity of experiments" he produced, and never doubted but that science in general, and English science in particular, benefited from his example as well as from his discoveries.

In many ways, English science as typified by the Royal Society remained a puzzle to Continental scientists, who could not believe that the methods the English preached were those which produced their undoubted results. Nor were these methods easy to follow. As noted above, not all English virtuosi succeeded in rising above a naive empiricism, try as they would. The English, on the whole, accepted the Royal Society's method as a model, even if they could not always emulate it. When provincial societies were founded—the Oxford Philosophical Society and the Dublin Philosophical Society in 1683, and a host of others in the eighteenth century, their avowed aim was to produce a copy of the Royal Society to the best of their ability, and those members who were Fellows strove diligently to keep up the standards, with varying success. But even within England, and certainly abroad, the role of empiricism in science remained a debatable one in the seventeenth century precisely because empiricists and rationalists had the same overt goals, and therefore could by no means always understand the need for agreement on methods. It was not until the next century that their aims diverged, that science and philosophy drifted ever farther apart, and that Continental mathematical physics acquired a sophistication that simple English empiricism could not achieve. The grounds of debate in 1800 were very different indeed from those of a century earlier.

References

1. Letter of 18 February 1662, Letter 262 of *The Correspondence of Henry Oldenburg* (eds. A. R. Hall and M. B. Hall), **ii** (Madison and London, 1966), 25–9. Hereafter referred to as *Correspondence*. Vols. **i–ix** were published by the University of Wisconsin Press; vol. **x** and subsequent volumes are to be published by Mansells.
2. Letter 302 of 25 December 1663; *Correspondence*, **ii**, 136–9.
3. "Dr. Wallis's Account of some passages of his own life", written in January 1696–7.
4. *A Defence of the Royal Society, an Answer to the Cavils of Dr. William Holder* (1678). Wallis's view that the Society took its first beginnings from meetings in London, not Oxford, was frequently stated by him in correspondence with Oldenburg and others in the 1660s and early 1670s, in passing comment.
5. See Thomas Birch (1755–6), *History of the Royal Society*, London, **i**, 88.
6. Letter 268; *Correspondence*, **ii**, 45–8.
7. Letter 291 (*Correspondence*, **ii**, 110–11); see the minutes for 23 September 1663 in Birch, *History*, **i**.
8. *Correspondence*, **viii** (1971), Letter 1858 (3 January 1671–2), pp. 448–51.
9. See Birch, *History*, **iii**, 10.
10. See his letter of 18 June 1673 (Letter 2254) and for Oldenburg's reply, Letter 2255, both in *Correspondence*, **x** (1975).
11. Letter 2407 of 15 December 1673 in *Correspondence*, **x**.
12. See Birch, *History*, **iii**, 122. Sand's reply is Letter 2449 of 27 February 1673–4 in *Correspondence*, **x**.

13. Birch, *History*, **iv**, 326 and ref. 9.
14. Letter 2404 of 12 October 1673 in *Correspondence*, **x**.
15. Letter 2422 of 14 January 1673–4 in *Correspondence*, **x**.
16. Minutes of the meeting of 11 December 1673.
17. Letter 2372 of 26 October 1673 in *Correspondence*, **x**.
18. Letter 2453 of 11 March 1673–4 in *Correspondence*, **x**.
19. Printed in C. R. Weld (1848), *A History of the Royal Society*, London, **i**, 146–8.
20. Chapter IV, Paragraph V, printed in Weld, *History*, **ii**, 526–27.
21. Letter 1240 of 12 July 1669; *Correspondence*, **vi** (1969), 108.
22. John Wallis (6 Aug. 1666), "An Essay of Dr. John Wallis, exhibiting his Hypothesis about the Flux and Reflux of the Sea", *Phil. Trans.* no. 16, 263–89. This was discussed by many, including Childry, Colepresse, Hyrne, Jessop, all virtuosi, in the succeeding years.
23. Letter 2445 of 20 February 1673–4. Correspondence, **x**.
24. In 1675, upon the presentation of his second paper on light and colours; see Birch, **iii**, 249, and *The Correspondence of Isaac Newton*, **i** (ed. H. W. Turnbull), Cambridge, 1959, p. 363.
25. See Birch, *History*, **i**, 43, 47–9 and, in general, Albert Van Helden, "Christopher Wren's *De Corpore Saturni*", *Notes and Records of the Royal Society*, **xxiii** (1968), 213–29.
26. See *Oeuvres, Complètes de Christiaan Huygens*, **iii**, 283, 321–22, 385–86 and Birch, *History*, **i**, 49.
27. The subject is discussed at length in A. Van Helden (1970), *The Study of Saturn's Rings 1610–75*, University of London Ph.D. thesis. There are some remarks in the Society's minutes for 1671, and more in its correspondence. See *Correspondence*, **vii** (1970) and **viii**, index s.v. Astronomy.
28. Letter 607, 6 February 1666–7; *Correspondence*, **iii** (1966), 337–39.
29. See Sluse's Letter 639 (*Correspondence*, **iii**, 412) and Oldenburg's Letters 644 and 644a (*Correspondence*, **iii**, 430 and 433).
30. In Letter 1284 (14 September 1669), transmitting information given him by Collins in Letter 1283 (*Correspondence*, **vi**, 227–8, 233).
31. Letter 1486 of 13 July 1670, *Correspondence*, **vii**, 64–8.
32. See Letter 1644 of 1 March 1670–71, *Correspondence*, **vii**, 485–92. The letter and hypothesis were presented to the Society on 23 March (Birch, *History*, **ii**, 475).
33. See Birch, *History*, **ii**, 477 (meeting of 20 April 1671). For Wallis's report (dated 7 April 1671), see Letter 1673, *Correspondence*, **vii**, 559–65; it was printed in *Phil. Trans.*, no. 74 (14 August 1671), 2227–30.
34. Birch, *History*, **ii**, 481 (meeting of 11 May 1671).
35. Birch, *History*, **ii**, 482 (meeting of 25 May 1671).
36. Letter 1713 of 2 June 1671; *Correspondence*, **viii**, 72–4. For the exchanges between Wallis and his former pupil Neile, see *Correspondence*, **v** (1968) and **vii**.
37. Letter 1724 of 12 June 1671, *Correspondence*, **viii**, 94–104.
38. Letter 1767 of 10 August 1671, *Correspondence*, **viii**, 191–3.
39. For Hooke's proposal see the minutes of the meeting of 22 October 1668. For the outcome see *Correspondence*, **v**, 103–4.
40. Letter 1794 of 10 October 1671, *Correspondence*, **viii**, 281, 283.
41. See *Correspondence*, **ix** (1973).
42. Cocherel wrote first at length about his longitude methods and lunar tables in Letter 1297 of 5 October 1669 (*Correspondence*, **vi**, 262–67). There is no record of Oldenburg's reply. Cocherel's Letters 2293 (of 8 August 1673) with more lunar tables (2293a) and 2385 (of 12 November 1673) were referred to Flamsteed, who reported unfavourably.

43. Letter 1995 of 12 June 1672, *Correspondence*, **ix**, 103–4. For the minutes see Birch, *History*, **iii**.

44. This is curious; Oldenburg sent a copy to Richard Towneley who copied it, and sent the first copy to Flamsteed, who also copied it before returning it to Oldenburg. (See *Correspondence*, **ix**, 204, 212, 226.) It may be that one of these copies still exists, but has not been traced.

45. Letters 2471 and 2472 of 30 March 1674, *Correspondence*, **x**. Oldenburg also sent a copy of Huygens (Letter 2470). Hooke had first described his method at a meeting of the Society on 28 July 1670, when "He was desired to prosecute carefully this observation, so important to determine the controversy concerning the motion of the earth".

46. Letter 2490 of 5 May 1674 (*Correspondence*, **xi**); the extract is printed in *Phil. Trans.* no. **105** (20 July 1674), 90.

47. Letter 2505 of 18 June 1675, *Correspondence*, **xi**.

48. Letter 2483 of 7 April 1674, *Correspondence*, **x**.

49. See also minutes for 21 January 1674/5. His observations were not further reported to the Society at this time, but became the Cutlerian Lecture *De potentia restitutiva* published in 1678.

50. *The Correspondence of Newton*, **i**, 362–86.

51. *The Correspondence of Newton*, **i**, 92–102. This is, of course, the first paper on light and colours.

52. Letter of 8 February; *The Correspondence of Newton*, **i**, 107. Oldenburg's letters show that the minutes normally contain less of the Society's enthusiasm for letters read before it than he was instructed to convey to their authors. But this reaction is unusually warm. The choice of Boyle was not empty; in 1744 a copy of Hooke's comments (10 pp. folio) was in Boyle's papers, together with Newton's two papers (82 pp.).

53. *The Correspondence of Newton*, **i**, 110–24 (also printed in Birch, **iii**, 10–15).

54. *The Correspondence of Newton*, **i**, 119.

55. *The Correspondence of Newton*, **i**, 171–88.

56. *The Correspondence of Newton*, **i**, 198–203.

57. *The Correspondence of Newton*, **i**, 362–86.

58. *The Correspondence of Newton*, **i**, 360–61.

59. *The Correspondence of Newton*, **i**, 412–3.

60. *The Correspondence of Newton*, **i**, 416–7.

61. On 6 February 1675–6.

62. *The Works of the Honourable Robert Boyle* (ed. Thomas Birch), 6 vols., London 1772, vol. **i**, 1.

63. Birch, *Boyle*, **i**, 354–9.

64. See *Correspondence, passim*.

65. Letter 2497 of 22 May 1674; *Correspondence*, **xi**.

66. For a more detailed discussion of this point, see A. Rupert and Marie Boas Hall, "Philosophy and Natural Philosophy: Boyle and Spinoza", *Mélanges Alexandre Koyré*, Paris, 1964, **ii**, 241–56.

67. Letter 244, Spinoza to Oldenburg, April 1662; *Correspondence*, **i** (1965), 462, 463. For the whole exchange see also *Correspondence*, **ii**.

68. "Nouveau essais sur l'entendement", *Die philosophischen Schriften von Gottfried Wilhelm Leibniz* (Berlin, 1882, Hildesheim, 1960), **v**, 436–7.

5. Science and Religion in the Seventeenth Century

P. M. RATTANSI
(*University College London*)

The image of Galileo before the Inquisition haunts any attempt to discuss the encounter of science and religion in seventeenth-century Europe. Haunts...but perhaps also distorts. For a long time that tragic event has been seen as embodying a remorseless logic. The new cosmology was not a purely technical matter for "mathematicians" to argue about, as Copernicus had insisted.[1] It constituted a dire threat to established churches. In subverting Aristotle, it brought into question the subtle conceptual apparatus the Church had created from his teachings over the centuries.[2] Even Protestants had reluctantly to admit its utility and, indeed, indispensability for articulating and resolving doctrinal problems soon after the Reformation.[3] The nature of the challenge was brutally clear in Galileo's "*Letter to the Grand Duchess Christina*" (1615). Reversing a traditional attitude, Galileo asked the theologian to bow to the authority of the natural philosopher whenever interpreting a Scriptural text which referred to the world of nature.[4] The response of the Church was predictable and inevitable.

It is a plausible picture. But we know now that relations between science and religion were far more complex than any such simple confrontation. For one thing, the new intellectual stirring of which interest in the "new science" was a manifestation extended into the ranks of the church itself. The Cardinal Barberini, who, as Pope Urban VIII, ordered Galileo's return to Rome in chains, if necessary, to answer charges before the Inquisition, had once proclaimed himself one of the "Galileisti" and celebrated Galileo's telescopic discoveries in Latin verse.[5] His favour and protection continued long after his ascension to the papal throne. Some of the leading pioneers of the "new philosophy" were in orders. Mersenne was a Minim Friar; Gassendi an ordained priest.[6] Observational astronomers of the first rank came from the Jesuits.[7]

Nor can the notion of a simple conflict be saved by contrasting a repressive Catholicism, fearful of scientific advance, with a tolerant or even sympathetic Protestantism. Pre-Tridentine Catholicism, at least, permitted a far more liberal interpretation of scriptural texts thought to conflict with scientific theories or discoveries, than a Protestantism which initially insisted on the supreme authority of the "naked word" of Scripture.[8]

Indeed, the judgement of an established church about the gravity of the threat it apprehended at any given moment from the new cosmology and the new science, and its own course of action, depended on many local and specific circumstances. Any close examination of the Galileo trial amply demonstrates that, and the deficiency of the simple view, whether adopted by a Voltaire or a Bertolt Brecht.[9]

The Church reserved the supreme penalty for the propagation of doctrines it regarded as clearly anti-Christian. Bruno was burnt at the stake in Rome in 1600 for preaching a return to an ancient Hermetic cosmic religion of which Christianity was a corrupt version; Vanini at Toulouse in 1619 for a pure-Aristotelian naturalism and mortalism.[10] There were no "martyrs for science" in the later seventeenth century. But Hobbes and Spinoza fell foul of established churches for the anti-Christian ways in which they were felt to have developed the new Mechanical Philosophy. Hobbes fashioned it into a new materialism and mortalism, Spinoza into a monistic and pantheistic world-view.[11]

To give up the assumption that Galileo was a secret agnostic or atheist is not the same as leaping to the other extreme. "Sincère mais banale" seems an apt description of his religious faith.[12] It would be a mistake, however, to regard that attitude as typical of seventeenth-century "scientists". Two sorts of confusion probably underlie the wide prevalence of that view. One confounds anti-clericalism, pervasive among intellectuals, with anti-religious opinions. The other confuses scepticism about the rational demonstrability of religious dogma with a lack of religious faith.

Certainly, the scepticism which is such a striking feature of European intellectual life by the late sixteenth century, constituted an important problem for pioneers of a "New Science" in the first half of the next century. Francis Bacon believed that a scepticism which denied rational certainty in all provinces of human knowledge constituted an insuperable barrier to the advancement of learning.[13] Both Mersenne and Descartes were convinced that a general scepticism and the discrediting of scholastic Aristotelianism had left the way open for currents inimical to Christian religion. The most dangerous ones were Aristotelian naturalism on the one hand and Hermeticism on the other. Each, as Lenoble has demonstrated in his well-known study, was regarded by Mersenne as destructive of the Christian distinction between natural law and miracle, and with it of the foundations of Christianity.[14]

Strategies to counter the dangerous consequences thought to flow from a pervasive scepticism varied from one thinker to another. By the time he came to write his *Le Monde* (1629–33), Descartes believed he had discovered a unified method to lay the foundations of certainty in religion and in natural philosophy.[15] Others had abandoned the hope of such intellectual unity and drew the bounds of natural theology far more narrowly. Received human knowledge was a maze of uncertainty. The human mind need not, however, despair of attaining certainty in its knowledge of nature—but only if it ceased to rely upon corrupted reason and turned to the methodical processing of sense-experience. While such knowledge brought proof of the existence of a divine intelligence, it could tell us little more about its nature or attributes.

That was notably the view of Francis Bacon. Man had access to two kinds of knowledge, "the one informed by the light of nature, the other inspired by divine revelation".[16] They were not to be confused. Sense could discover natural things, but would only darken divine things. Insofar as knowledge of nature brought evidence of providential design, it provided a support for faith, but only of an indirect character: "The bounds of this knowledge [natural theology] are, that it sufficeth to convince atheism, but not to inform religion."[17]

Two features of Bacon's approach are especially noteworthy. One is the influence of a late-medieval philosophical tradition, influential in England, which had sapped the credibility of rational theology and which had nourished, and in turn derived support from, a major current of Reformation thought.[18] The other is the way in which a method supposedly resting entirely on observation and induction presupposed such a distinction if it was to be compatible with religion. Otherwise it risked drawing upon itself the hostility of the church to an epistemology identified historically with the ancient Epicureans.[19]

A contrast can certainly be made between the unified rational method Descartes hoped would furnish evidence for fundamental religious principles as well as indubitable starting points for natural philosophy, on the one hand, and the alternative view that the certainty of faith, though buttressed by evidence of design in nature, was founded on revelation, while that of scientific knowledge rested on induction from sense-experience. But seventeenth-century science did not reflect any straightforward conflict between the "rationalists" and "empiricists" beloved of many historians of philosophy. Descartes consigned to observation and experiment the task of deciding between rival hypotheses equally compatible with a basically mechanistic approach to nature. And the empiricist programme of Bacon became really fruitful only when harnessed to a Mechanical or Corpuscular Philosophy.[20]

Harmony and even active co-operation between science and the dominant belief-system in seventeenth-century Europe were not then merely the un-intended consequences of an elective affinity like that which has been suggested between Puritanism and science. Some of the leading pioneers of a new science, like Mersenne and Descartes were actively concerned to furnish new weapons to defend religion at a time when the old arguments seemed to have been discredited.[21] The problem was seen as one of fundamentals, since those arguments had ceased to carry conviction because the old conceptual structure in terms of which they did so had lost its intellectual authority.

There were others who renounced any ambition of re-erecting an integrated theological-metaphysical-scientific system in the spirit of St Thomas. They separated the province of science from that of faith by defining the different sources of valid knowledge. Bacon argued on the whole for such a separation.[22] Pascal is an interesting example of a thinker who made a similar division, but who came increasingly to depreciate and ultimately to abandon scientific pursuits as distractions from the religious quest.[23] A tendency to contrast scientific and religious truths in order to emphasise the far greater degree of certainty of the latter is characteristic of Robert Boyle. It never persuaded him to abandon his massive experimental labours, but gives a tension to his thought that we shall miss if we dismiss the theological tracts which occupy much of his collected works as irritating distractions from his solid scientific investigations.[24]

Is it possible to discern "national patterns" in the relation between science and religion in seventeenth-century Europe? At one level, that is obviously so. The Galileo-trial may be seen as an example of a general repressiveness in Italian intellectual life in the interests of religious uniformity which has been traced as already evident during the pontificate of Clement VIII, in the last decade of the sixteenth century. Patrizi's *nova philosophia* and Telesio's *de rerum natura* were placed on the Index, as were the entire works of Bruno and Campanella. Investigations were begun into the works of della Porta and Cremonini, and the early years of the next century saw the beginning of the long imprisonment of Campanella and the burning of Bruno.[25] In France, by contrast, while "libertine" doctrines or theological innovations were put down by a royal absolutism when seen as symptoms of civil revolt, a vigorous intellectual life flourished during the century. The Sorbonne periodically reaffirmed the supremacy of scholastic-Aristotelian authority but was never able to enforce the penalties it threatened to dissent.[26]

Even where the church could not enforce such measures, its condemnation of new doctrines was significant since it controlled higher education. In England, the church never took up an official position against the new cosmology or the new science; but Archbishop Laud, in his revised Statutes for Oxford of 1634, took the occasion to reaffirm the supreme authority of Aristotle.[27] When Newton came to Cambridge in the early 1660s, he was required to study natural philosophy from German-Protestant neo-scholastic compilations.[28]

The study of national contexts furnishes clues to attitudes to emergent science in relation to attitudes to intellectual innovations in general. Are there other ways in which such a study can illuminate the development of natural philosophy in the seventeenth century? That question may be answered by examining the suggestion that, in order fully to understand the thought of Francis Bacon and the curious strength of "Baconianism" in seventeenth-century English thought, we must take account of a particular English philosophical and theological tradition reaching back to the critical and sceptical tendencies of the 14th and 15th centuries, as modified in the 16th. By the 18th century the French *philosophes* had made a cult of an English tradition which they saw stretching from Bacon to Newton and Locke, and marked by a freedom from *a priori* assumptions and system-building ambitions, and by reliance on observation and experiment as the basis for scientific theories.[29]

It is true that the influence of Bacon began to wax in England soon after his own death in a way that suggests that his doctrines were in harmony with important values in the English intellectual environment. In the disturbed decades from 1640 to 1660, Bacon's authority was invoked both by the radical university reformers and those they battled with about curricular reforms—the members of the newly-installed Oxford group of natural philosophers. Even those less interested in Bacon's scientific programme cited his *Advancement of Learning* when defending the universities against extreme sectaries who called for abolition of universities as the idolatrous high places condemned in scripture.[30]

The ideas of Bacon influenced the Oxford group in some fundamental ways. Their early projects were in accord with the design he had sketched. They wished to learn from the crafts. They stressed the ultimate aim of enriching human life with inventions and discoveries. In their pronouncements on scientific method, they had already begun to contrast their cautious "Baconian" empiricism against premature continental and principally Cartesian system-building. Certainly, it is legitimate to compare Boyle's insistence on the rigorous experimental testing of theories with the plausible explanations Descartes gave for a large class of phenomena.[31]

Nevertheless, the Philosophy of Science should serve to remind us that in considering pronouncements on scientific method we have entered a realm haunted by myths. It was the new mechanical philosophy, most systematically formulated by Descartes and Gassendi, which caught the imagination and enthusiasm of the young natural philosophers in the London and Oxford groups, and furnished them with a programme for research. But that new framework for science was a metaphysical construct, a new conceptual system, with a necessarily problematic relation to "observation and experiment", or "ocular demonstration". It could not be a grand theory arrived at as the culmination of a programme of induction from particulars. Lamenting the death of Robert Boyle in 1691, Leibniz and Huygens commented on what they considered the extraordinary fact that Boyle had devoted his vast labours to proving something he had begun by assuming, that is, the mechanical explanation of nature.[32] That comment may seem unfair, because Boyle's experiments were often designed to provide the experimental basis for a choice between theories equally compatible with a mechanical explanation—but in that case the possibility of making a very sharp distinction between the procedures advocated by Descartes and those adopted by the English movement, or of making Bacon or even Gilbert the "father of the New Philosophy" as Wren did in a famous oration became far less convincing.[33]

In studying the development of science in national contexts, we should therefore be sensitive to this mythical element in the self-image of scientists and scientific movements or institutions. But once we do so, we can choose to regard them as significant clues to a sociological dimension of the history of science: to values of crucial importance in a given society.

We should also take full account of the existence of traditions other than what appears at a given time to be the dominant "national" tradition. When considering the formulation of "Baconianism" as the ideology of the English scientific movement after about 1640, for example, we must take account of two divergent currents of thought.

One stems from the Cambridge Platonists. Unlike the members of the Oxford group, they initially received Cartesian doctrines with enormous enthusiasm. What impressed them was its philosophical unity, which provided a rational demonstration of the existence of spiritual entities as part of the same enterprise which yielded an "intelligible" mechanical account of nature. They were contemptuous of the induction and utilitarianism of Bacon. They pointed to the radical inadequacy of a naive empiricism which was as powerless to furnish firm starting-points for scientific inquiry as it confessedly was to lend support to religious principles.[34]

The other major exception was Thomas Hobbes. The Cambridge thinkers saw him as exemplifying the dangers that lurked in a scientific system not indissolubly linked to the Christian theological and philosophical system. A pioneer of the mechanical philosophy, and one drawing upon the English nominalist tradition, Hobbes was nevertheless scathing in his attacks on the "Baconian" empiricism of the Royal Society and of Boyle in particular.[35] He aimed at intellectual economy, without the insoluble problems of dualism, by abolishing the notion of spirit. The mechanical philosophy was to be applied not only to the world of nature, but to human minds in psychology, and to human wills in society and polity.

Henry More, the Cambridge Platonist, was to be elected a Fellow of the Royal Society; but he made no secret of the fact that he thought their most useful activity would be the collection of ghost-stories to provide empirical confirmation of the existence of spiritual entities capable of influencing matter, as against Hobbist materialism. And the works he wrote after his final disenchantment with Descartes, arguing that there were no natural phenomena capable of a completely mechanical explanation, may seem to us to have contributed little to seventeenth-century science. However, we are beginning to see that much that is baffling about the thought of Isaac Newton—when we see him in the light of the dominant modified "Baconian" tradition—becomes clearer when we take fuller account of the influence of the Cambridge Platonists upon him.[36]

The significance of Hobbes for the development of natural philosophy was something that already in the 1650s and 1660s prominent members of the Oxford group made it their business to deny. In one work after another, they attacked his scientific competence in order to avert the danger, which they considered grave and real, of a popular identification between the mechanical philosophy and mortalism and materialism. Only now is it beginning to be reassessed and to be found more solid and important than hitherto thought.[37]

I have tried to view the problem of the relation between religion and science in 17th century England as a part of the historical problem of defining the relation between science and the dominant belief and value-systems in particular European societies during the period. In doing so, however, we must be careful to take account of a variety of divergent traditions of natural philosophy and religion. We must also recognise the mythical element in the ideal image of science held and propagated by natural philosophers or scientific groups, and explore the ways in which they point to significant aspects of the belief- and value-system of the society.

References

1. "Mathematics are for mathematicians", Preface to *De Revolutionibus Orbium Caelistium* (1543), tr. by J. F. Dobson and S. Brodetsky, *Occasional Notes of the Royal Astronomical Society* **ii** (1947), 6.

2. For example, Bertrand Russell (1947), *Religion and Science* (1935), Oxford, Ch. 3, "The Copernican Revolution", pp. 19–48.

3. Discussion and bibliography in E. Lewalter (1935), *Spanisch-Jesuitische u. Deutsch-Lutherische Metaphysik des 17. Jahrhunderts*, Hamburg.

4. "Letter... Concerning the Use of Biblical Quotations in Matters of Science", Eng. tr. in Stillman Drake (1957), *Discoveries and Opinions of Galileo*, New York, 172–216.

5. Maffeo Barberini, "Adulatio perniciosa", cited in G. di Santillana (1961), *The Crime of Galileo* (1955), London, 156, ref. 3.

6. R. Lenoble (1971), *Mersenne ou la naissance du mécanisme*, Paris, 2nd ed.; B. Rochot (1944), *Les travaux de Gassendi sur Epicure et sur l'atomisme*, Paris.

7. F. Russo, "Catholicism, Protestantism, and the Development of Science in the 16th and 17th Centuries", *J. World Hist.* **iii** (eds. G. S. Métraux and F. Crouzet), reprinted in *The Evolution of Science*, New York, 1963, 291–320.

8. O. Pfleiderer (1891), *Die Entwicklung der Protestantische Theologie*, Frankfurt; cf. R. Hooykaas (1972), *Religion and the Rise of Modern Science*, Edinburgh, 98–160.

9. See Voltaire on Galileo as "le restaurateur et la victime de la raison en Italie...", in *Oeuvres complètes*, Paris, 1877 **vi**, 335; Bertolt Brecht (1958), *Leben des Galilei, Schauspiel*, Berlin.

10. F. A. Yates (1964), *Giordano Bruno and the Hermetic Tradition*, London; on Vanini,

J. A. Charbonnel (1919), *La pensée italienne au XVIᶜ siècle et la courant libertin*, Paris, 354–84.

11. F. Brandt (1928), *Thomas Hobbes' Mechanical Conception of Nature*, Eng. tr., London; H. A. Wolfson (1961), *The Philosophy of Spinoza* (1934), Cleveland-New York.

12. Discussed in G. Spini, "The Rationale of Galileo's Religiousness", in *Galileo Reappraised* (ed. C. L. Golino), Berkeley and Los Angeles, 1966, 44–66.

13. "The Great Instauration", The Plan of the Work, p. 253, "Novum Organum", Author's preface, p. 256, in *The Philosophical Works of Francis Bacon* (ed. J. M. Robertson), London, 1905; R. H. Popkin (1964), *A History of Scepticism*, New York.

14. Ref. 6, 373–82.

15. N. Kemp Smith (1963), *New Studies in the Philosophy of Descartes*, London, 24–6.

16. *De augmentis scientiarum*, Bk. 3, Ch. 2, 456.

17. See also "Of the Advancement of Learning", ref. 16, 167–70.

18. M. H. Carré (1949), *Phases of Thought in England*, Oxford, esp. pp. 224–79.

19. Discussed by E. Gilson (1961), *The Christian Philosophy of St. Augustine*, Eng. tr., London, ref. 21, 280–1.

20. Compare Marie Boas, "The Establishment of the Mechanical Philosophy", *Osiris* (1952), 412–541, and R. H. Kargon (1966), *Atomism in England from Hariot to Newton*, Oxford.

21. Kemp Smith, ref. 15, 163–220.

22. Kuno Fischer (1857), *Francis Bacon of Verulam*, Eng. tr., London, Ch. 10–11, 290–371.

23. Pascal, see esp., "Préface sur la Traité du vide", 772–785, in *Oeuvres complètes* (1970) (ed. J. Mesnard), Paris i, 772–85.

24. "The Excellency of Theology compared with Natural Philosophy", 1–66, in *Works* (1772) (ed. T. Birch); the background is discussed by L. I. Bredvold (1956), "The Crisis of the New Science", in *The Intellectual Milieu of John Dryden* (1934), Ann Arbor, 44–72.

25. Luigi Firpo (1950), "Filosofia italiana e contrariforma", Eng. tr. in *The Late Italian Renaissance, 1525–1630* (ed. E. Cochrane), London (1970), 266–84.

26. R. Mousnier (1961), *Les XVIᶜ et XVIIᶜ siècles*, Paris, 277.

27. C. H. Mallett (1924), *A History of the University of Oxford* ii, London, 322.

28. R. S. Westfall (1962), "The Foundations of Newton's Philosophy of Nature", *Br. J. Hist. Sc.* i, 171–82.

29. G. Gusdorf (1971), *Les principes de la pensée au siècle des lumières*, Paris, 151–292.

30. George Williamson (1960), "Richard Whitlock, Learning's Apologist", in *Seventeenth Century Contexts*, London, 178–201.

31. See citations in G. H. Turnbull (1953), "Samuel Hartlib's Influence on the Early History of the Royal Society", *Notes and Records Roy. Soc.* x, 101–30.

32. *Oeuvres complètes de C. Huygens*, La Haye, 1905 x, 239, 269; also Leibniz, "Noveau essais" (1704), in *Oeuvres philosophiques* (ed. P. Janet), Paris (1866) i, 483–4.

33. In the English version of Gresham inaugural lecture, as given in S. Wren (1740), *Parentalia, or, Memoirs of the Family of the Wrens*, London, 204; but not in the Latin version, in J. Ward (1730), *The Lives of the Professors of Gresham College*, London, 29–37.

34. E. Cassirer (1953), *The Platonic Renaissance in England*, Eng. tr., London, 42–85; C. E. Raven (1953), "Cudworth and the Age of Genius", in *Natural Religion and Christian Theology*, Cambridge i, 99–124.

35. R. Boyle (1662), "An Examen of Mr. T. Hobbes his Dialogus Physicus de Natura Aeris", in *Works* i, 186–242.

36. P. M. Rattansi, "Some evaluations of reason in sixteenth and seventeenth century natural philosophy", in *Changing Perspectives in the History of Science, Essays in Honour of Joseph Needham* (eds. M. Teich and R. M. Young), London (1973), 148–66.
37. J. F. Scott (1938), *The Mathematical Work of John Wallis*, London, lists some of the attacks; also S. I. Mintz (1962), *The Hunting of Leviathan*, Cambridge; for a re-evaluation of Hobbes' contribution to optics: A. G. Shapiro (1971), "Kinematic Optics: A Study of the Wave Theory of Light in the Seventeenth Century", *Arch. Hist. Ex. Scs.* **xi**, 134–266.

6. The Growth of Science in the Netherlands in the Seventeenth and Early Eighteenth Centuries

W. D. HACKMANN

(*University of Oxford*)

The main purpose of this paper is to demonstrate that the development of science in the Dutch Republic during the seventeenth and early eighteenth centuries was influenced by a number of complex interrelated factors, and to analyse what these factors were. Some of these, such as changes in methodology, the secularisation of knowledge and the break-away from scholasticism and, in the early eighteenth century, the growing interest in natural theology, were international, but there exist also several specific Dutch factors such as the political situation and organisation of the newly founded Dutch Republic, the attitudes of the politicians and clergy to science in particular and education in general, and the structure of the recently established universities. This is, of course, a very large subject, so what follows is only the merest of outlines, which hopefully also indicates the richness of the material that exists in the Netherlands on this topic.[1]

The epic struggle against the Spanish king Phillip II by the inhabitants of the seventeen provinces in the north which made up his Bourgondian-Habsburg possessions, and from which later emerged the modern states of Belgium and the Netherlands, has been graphically recounted by such historians as Motley, Geyl and Huizinga.[2] The revolution was the result of a mixture of religious, economic and political factors. Thus, both the suppression of the Calvinist religion by a distant Catholic king and his administration, and the extra taxes levied for the upkeep of his armies, were contributary causes. Although Holland, which was the wealthiest and most powerful province of the Union, was free from Spanish domination by about 1580, and the newly formed Dutch Republic was recognised by England and France in 1596, the struggle continued until the end of the Eighty Years War in 1648. Throughout this period, the leaders of the new state had to exercise all their diplomatic skills so as not to fall prey either to its powerful allies or its enemies.

The Dutch state that emerged consisted really of a "commonwealth", a loose federation of important towns and provinces governed by the Stadtholder,

89

the Pensionary or Chief-magistrate, the committee of regents from each one of the seven provinces of the Union known as the States-General which usually sat at The Hague, and at a local level by the regents making up the town councils, and by the magistracy. These regents were members of the rich merchant class. The landed gentry had little power in the new Republic. During these early years, there existed a great deal of social mobility, and anyone with a shrewd sense of business could gain entry into the monied upper class and the magistracy. By the second half of the seventeenth century, however, the regent class started to become more rigid socially. Its members began to invest in land and buy up titles instead of continuing the expansion in business and banking. To some extent this change was caused by growing international competition, especially from England, which put a halt to Dutch colonial expansion and contracted its overseas trading. The financial burden during the second half of the seventeenth century, caused by the three short wars with England over trade and fishing, also contributed to the Republic's gradual economic decline. These wars demonstrated very clearly the vulnerability of a small nation living by overseas trade to foreign agression. By the middle of the seventeenth century the Republic's most creative period was over.

The educational reform that took place in the Republic in the early seventeenth century and its contemporary international reputation of liberalism to new philosophical ideas, was chiefly as a result of the long fight for independence and the resulting peculiar decentralised structure of government.[3] Other important factors were the existence of a large immigrant population more intent on trade than on the persecution of novel ideas, and of a fairly weak national church which was not directly involved with governing the state. Although the Reformed Church was largely successful in its attempts to become the state church, it was governed by church synods at both a provincial and a national level, thus, a structure similar to the secular government, and even though the Calvinists were by far the strongest sect, the Reformed Church was weakened by internal factions such as Remonstrants, Contra-Remonstrants, Socinians and a host of others. Conflicts which arose between the synods and the more liberal secular government were usually resolved by compromise after endless discussions in the various committees. A good example of this process is the way in which Descartes's scientific and philosophical ideas were assimilated into the Netherlands.

During the 1640s the conflict over Cartesianism between the theology and the philosophy faculties of the Dutch universities was steadily increasing, especially at Leyden where the disputations between pro- and anti-Cartesians often ended in physical brawls. In 1651 the Stadtholder Lodewijk Hendrik asked several Dutch universities for their opinion about this matter, and the reply he received from Harderwijk sums up the general feeling about Descartes: they had much praise for his mathematics and physics but not for his medical theories, while his metaphysics was thought to be pernicious to theology.[4] The problems caused by his ideas at Leyden were discussed both by the synods and the regents at local and provincial level, but no solution could be found which would satisfy both factions. Finally, following a suggestion made by the Pensionary Johan de Witt, the States of Holland in The Hague in 1656 formulated an oath that had to be taken both by the theology and philosophy professors. In essence this stated that the philosophy faculty could not teach Cartesianism, and that neither faculty could encroach on each others subjects. The resolution of the States added: "Although much has been said about the generous attitude of the States towards the liberty which is necessary for philosophizing, nobody must imagine that because of it the public and accepted manner of teaching may be dispensed with."[5]

After some hesitation, the philosophy faculty accepted the taking of this oath, while the professors of theology, of course, took it much more readily, and the trouble at Leyden was averted. This oath, however, should be seen more as a psychological sop to the clergy than as a deterrent to the ideas of Descartes which continued to be taught in the Republic. Furthermore, this oath was only applicable at Leyden as the other provinces resolved this problem in their own way. Thus, in 1656 an academic law was passed at the University of Harderwijk which prohibited the teaching of Cartesianism, and this may have been the reason why the pro-Cartesian professor of philosophy Cornelis van Thiell was temporarily suspended in 1657.[6] The only victim at Leyden was the aged professor Heidanus who was dismissed from his Chair in 1676 after he had written a pamphlet in defence of Descartes's metaphysics.[7] In Utrecht, however, there was considerably more trouble, but this was largely caused by the unstable political situation. After this city was taken back from the French in 1673, more than 120 liberals lost their posts in the city-government, and in the University, too, Chairs which were vacated in the faculties of theology and philosophy, were filled by the anti-Cartesian disciples of the theologian Voetius.[8] However, these were the last skirmishes in the Cartesian debate in the Dutch Republic. The conservative clergy started to attack Spinoza instead, while Cartesianism began to be watered down by the empiricism of Bacon and Boyle. These varied reactions to Cartesianism demonstrate the important consequences of the Dutch decentralised structure of government on the freedom of expression, as it allowed those, such as professors who held strong views, either conservative or liberal, to move to those districts where their ideas were more acceptable, and it also assisted in the freedom of the press, as books were usually suppressed only at a local level.

The population of the emerging Dutch Republic was an extraordinary mixture of nationalities. In about 1620, there were less than one and a half million inhabitants, but by the end of the eighteenth century this had increased to two millions, and a large proportion of these were immigrants. Thus, in Amsterdam alone in the 1620s, one-third of its 100 000 inhabitants were foreigners.[9] Most of these had fled from the intolerance existing in their own countries or were tempted by the wealth and trading possibilities in the new state. As Andrew Marvel wrote in 1653:

"Hence *Amsterdam-Turk-Christian-Pagan-Jew*,
Staple of Sects, and mint of Schism grew;
That *Bank of Conscience*, where not one so strange
Opinion but finds Credit and Exchange."[10]

Many of these immigrants were Huguenots from France, and by the early 1700s about 55 000 had settled mainly in the towns of the provinces of Holland and West Friesland. A large number of these became ministers and teachers both in secondary and higher education.[11] Well known amongst these were Papin and Jacques Bernard; the latter was appointed professor of natural philosophy and mathematics at Leyden in 1705.

It was on this multi-national society that the first Stadtholder of the new Republic, William of Orange, better known as William the Silent, tried to impress a national identity. He realised that one important way of achieving this aim was to create a system of higher education for Dutchmen.[12] This could furthermore supply the state with protestant teachers and ministers. He founded Leyden University in 1575 as a reward to the town for having successfully withstood a lengthy Spanish siege.[13] After William the Silent's assassination in 1584, his progressive educational policy was continued by his successors and by the States-General. The University of Franeker was founded in 1585, Harderwijk in 1600, Groningen in 1614 and Utrecht in 1636. Several "Illustrious Schools" were also established, such as Deventer in 1630, Amsterdam in 1632, Dordrecht and 's Hertogenbosch in 1636, Breda in 1646, Nijmegen in 1655 and Maastricht at about the same time.[14] These schools did not have the same prestige as the universities for they could not award degrees. In most cases they were preceded by a Latin School to which two additional classes were simply added for the teaching of the Liberal Arts, although some theology and medicine were also taught. These schools prepared the students for the three major faculties of medicine, theology and law taught at the universities. In general, the "Illustrious Schools" were not well endowed, and they had little influence on the development of philosophy and science in the Dutch Republic. Most of their professors were ministers who besides teaching had to continue with their religious duties.

The five universities, however, were far better endowed, especially as the government viewed them as a means of gaining international prestige for the new state. One way of achieving this was to tempt well-known foreign scholars by means of financial inducements such as very high salaries, and sometimes free housing and transport. Thus, during the seventeenth century, Groningen had a total of fifty-two professors, including thirty-four foreigners. Twenty-seven professors were of German origin, but others came from France,

Switzerland, Ireland and Scotland,[15] while during the same period about one-sixth of the teaching staff at Leyden and Utrecht were foreigners.[16] Well-known teachers attracted to Groningen were the Scot Macdowell,[17] the mathematician Johan Bernoulli from Basle,[18] and the moral philosopher Pasor from Ireland.[19] Two famous men who were tempted to Leyden were the botanist Jules Charles de l'Escluse, or Clusius, and the physician Archibald Pitcairn who resigned his post in a rather ill-mannered fashion in less than a year.[20] That well-known scholars were chosen with the intention of attracting students is demonstrated very clearly in the case of Clusius. When in 1593 this frail old man was appointed at a high salary to the Chair of Botany, he was exempted from teaching (at his own request), and even his duties at the botanical garden were taken over by an assistant, the botanist T. A. Cluyt.[21] The largest number of the early professors of medicine, botany and anatomy at Leyden had studied at Padua.[22]

This policy of attracting good teachers was very successful which is demonstrated by the large number of foreign students who studied at the Dutch universities during the seventeenth and early eighteenth centuries. Peacock has listed 4300 English-speaking students who went to Leyden during the period 1575–1835, but as indicated by the *Album studiosorum*, most of these were there during the seventeenth century.[23] Of these 2124 were medical students.[24] When Guthrie analysed these figures further in 1959, he discovered that on average one-quarter of these were Scottish students. This proportion increased to one-third during Boerhaave's period from 1701–38. Indeed, most of the early professors of the Medical School of Edinburgh University which was started in 1726, had studied for part of the time at Leyden.[25] The same was true of Edinburgh's first professor of botany, Charles Preston, and of chemistry, James Crawford.[26] In fact, the whole structure of teaching at Edinburgh bore strong affinities with Leyden. During the same period more Cambridge medical students went to Leyden than to any other foreign university.[27] However, as has been pointed out by Rook, a Leyden M.D. did not imply that the student had received a significant part of his training there. Indeed, he lists eleven different types of stay in Leyden. Some students obtained their M.D. after only a few days or weeks, others stayed for a number of years, while others again never took their doctorate at Leyden at all, some taking it on their return to Cambridge.[28] A number of these students would have left Cambridge during the 1630s because of the political and religious disturbances at the English universities, while others undoubtedly were attracted by the teaching of the professor of medicine Otto van Heurne who introduced clinical training in 1637 soon after it had been started at Utrecht.

The medical faculty at Leyden was extremely progressive, and both Sylvius de Le Boë and its most successful teacher, Herman Boerhaave, realised the importance not only of clinical teaching, but also of anatomy, physiology and chemistry. After the death of Boerhaave in 1738, the number of foreign medical students at Leyden declined very rapidly. Similar statistics can be given for the other Dutch universities, although there the number of English students was much smaller. Thus, at Groningen during the first seventy-five years from 1614 to 1689, there were 3348 Dutch and 2642 foreign students, but of these only eleven came from England and Scotland. For the period 1689 to 1808 the *Album studiosorum* lists 4371 Dutch and 1215 foreign students which include seventy-five from England and Scotland. Most of these were Germans and Hungarians,[29] and the same was true of the foreign students at Harderwijk.[30] Many of these came to be trained for the Calvinist ministry. Leyden had, however, the best international reputation, at first because of its progressive medical school, but after the appointment of De Volder in 1675, also for experimental philosophy. This latter reputation was further enhanced by the appointment of 's Gravesande in 1717 and Petrus van Musschenbroek in 1739.

As the Dutch universities were all city foundations they did not have the scholastic traditions of the older institutions in other countries. Their progressive attitude is shown by the relative ease with which scholastic Aristotelianism was overthrown in the philosophy faculties and new ideas were accepted, such as those of Pierre Ramus. In this way they were similar to the eighteenth-century English Dissenting Academies, especially in their interest in science.[31] As in other countries, however, many of the best-known Dutch seventeenth-century natural philosophers such as Christiaan Huygens, Antony van Leeuwenhoek and Jan van Swammerdam, were not members of a university. Nor were they a homogeneous group, but each one in his own way is a good example of this period. Thus, both Huygens and Swammerdam received the conventional university education, but broke away from the scholastic tradition and advanced their own topics through keen observation and experiments. Leeuwenhoek did not go to university but was educated for trade. He opened a draper's shop in Delft, the town of his birth, in 1654. Both his marriages (1654 and 1671) connected him with wealthy and influential local families; his first wife came from a draper's family and the second was the daughter of a clergyman. His financial position improved over the years and he was given several municipal appointments: chamberlain to the sheriff's office, municipal wine-gauger and general district superintendent, which demonstrate that he was respected by the local community. After he had become well known for his microscopical observations, most of these posts were treated as sinecures.[32] Leeuwenhoek only knew Dutch and his scientific reputation was solely based on his powers of observation, his common-sense and integrity. He would only accept those things which he had actually seen with his own eyes. During the early years, the more cultured scholars such as Constantijn Huygens regarded him with a certain amount of condescension, so that in a letter to his better-known son Christiaan he could refer to

Leeuwenhoek as "notre philosophe bourgeois à delf", writing further: "Vous voijes comme ce bon Leeuwenhoek ne se lasse pas de fouiller par tout où sa microscopie peut arriuer. si beaucoup d'autres plus sçauans vouloijent pendre la mesme peine, la decouuerte des belles choses iroit bientost plus loing."[33]

Christiaan Huygens and Leeuwenhoek, however, had one important thing in common that characterised the new approach towards scientific research. Both used the empirical techniques of the craftsmen and artisans as advocated by Francis Bacon and others. Both were also extremely practical and made their own research tools. Thus, they manufactured their own lenses and Christiaan devised his own clocks.[34] The technical expertise of the Dutch in shipbuilding, land-drainage, canal and dyke construction, are well known, but these topics were not taught at the universities. These skills, in common with other trades and crafts, were passed on by the training of apprentices, and advances were usually made by means of careful observations and trial-and-error. During the sixteenth century a movement started which suggested that the same techniques could be applied with benefit to scientific research. The French advocate for educational reform, Pierre Ramus, admitted freely that he had profited from instrument-makers, engineers and surveyors, and was proud of the fact that he knew all the workshops in Paris. Furthermore, both Bernard Palissy in France and Francis Bacon in England suggested that knowledge about nature could be advanced through empirical observation and experimentation.[35]

Simon Stevin at The Hague and Isaac Beeckman at Middelburg[36] were early exponents of this movement. Thus Beeckman for a long time continued his father's trade of candlemaking as this allowed him more free time for his scientific experiments,[37] and Stevin, who was appointed tutor in mathematics and natural philosophy and financial adviser to the Stadtholder Prince Maurice of Orange, attempted to improve the crafts by means of more systematic research. Connected to this move away from the scholastic tradition was the use of the vernacular in place of Latin. In the Dutch Republic Stevin was the extreme exponent of this. Not only did he realise that by writing in the vernacular, science was opened up to the craftsmen and others who had not received a formal university education, but he also argued that Dutch was the best language for scientific reasoning.[38] This latter somewhat eccentric view might in part be owing to his close connexion with Prince Maurice and the Dutch fight for freedom; the use of the vernacular assisted in establishing a national identity. The same kind of arguments were put forward by others, including in 1669 by the Dutch playwright Lodewijk Meyers who is better known abroad as the publisher of Spinoza's *Renatus des Cartes Principiorum Philosophia* (1663).[39] Latin, however, remained the academic language at the recently established Dutch universities, and this made these institutions much more accessible to foreign scholars.

A major factor governing the growth of science in the universities was the structure of the faculties. Until the reorganisation of university education in the Netherlands in 1815, there existed four faculties; the three major ones of theology, law and medicine, and the subordinate fourth faculty of the liberal arts which included philosophy and mathematics.[40] In theory, students could not enter one of the higher faculties until they had completed their courses in the arts faculty, but in practice many disregarded this rule, so that at Franeker, for instance, in 1663 the M.A. and doctorate degrees were combined to "L.A.M. et Phil. Dr."[41] There existed great flexibility in the courses which could be taken in the arts faculty, thus one could concentrate on the classical languages or on philosophy and mathematics. This flexibility probably contributed greatly to the number of Reformed Church ministers, lawyers and doctors who departed from university with an interest in science, and created the scientific milieu so apparent in the eighteenth century. Chemistry was not taught in the arts faculty, but became part of the medical curriculum because of its practical application in medicine and pharmacy.[42] Thus, the arts faculty was treated as a link between secondary education and the training in one of the three professional faculties. In this sense it performed the same function as the "Illustrious Schools".

This system, however, also had a drawback which did not benefit the subjects taught in the arts faculty. Its professors were considered to be of lower status, were paid less and often combined their teaching with other jobs. Especially in the early years, conflicts often arose when professors tried to gain Chairs in the higher faculties or increase their sphere of influence by combining different courses. Also, in times of economic depression, it was always the arts faculty which was worst hit. These factors resulted in a considerable movement of professors between subjects within the faculty. By the first quarter of the eighteenth century, however, owing to the tremendous increase in the popularity of natural philosophy, this topic had predominated over all the others.

The first professor of experimental philosophy at Leyden, Buchardus de Volder, introduced the earliest *Theatrum physicum* in 1675, after a visit to London during the previous year, where he was probably inspired by the instrument collection which he had seen at the Royal Society.[43] He was soon assisted by the anti-Cartesian professor of philosophy Wolferdus Senguerdius who taught a mishmash of conventional scholasticism and experimentalism, and had been undoubtedly appointed to counteract De Volder's Cartesian sympathies.[44]

Instruments were introduced at Groningen in 1669 by Johannes Bertling, the professor of logic and mathematics. These were used in his lectures on balistics, fortification, surveying and navigation.[45] Experimental apparatus was added to this teaching collection by Johan Bernoulli in 1697 after he had been appointed to the Chair of mathematics.[46] He only stayed there until

1705 when he returned to Basle to take over his deceased brother's mathematics professorship, and his instruments remained unused until the late 1720s. At Harderwijk some experimental philosophy lectures were given for the first time by the philosophy professor Adrianus Reeland in 1700 when he obtained an air pump and some other apparatus, but he left after only a few weeks to take up the professorship of oriental languages at Utrecht.[47] At Franeker, too, the first attempt to introduce experimental lecture-demonstrations was short-lived. Some instruments were purchased in 1701 by the professor of logic and physics, Ruardus Andala, but as a Cartesian he appears to have come into conflict with several of his colleagues so that he stopped these courses. In 1712 he was appointed to the Chair of theology.[48] The last university to introduce experimental physics was Utrecht in 1705. Here too the subject was taught by the philosophy professor, Joseph Serrurier, who continued until 1716 when the topic remained in abeyance until the appointment of Petrus van Musschenbroek in 1723. It was largely through the latter's efforts that this subject really began to flourish at Utrecht, and substantial sums were spent by the town-council on the purchasing of experimental apparatus.[49] Much of this can still be seen at the Utrecht University Museum.[50] In 1739, however, van Musschenbroek moved to Leyden where he remained until his death in 1761. Apart from Groningen, at all the other universities, experimental philosophy was introduced by ex-Leyden students who had followed the courses of De Volder. Until the 1730s, however, Leyden remained the most successful place for this subject. The other universities renewed their interest after Newtonian experimentalism had been firmly established by 's Gravesande and his successor Petrus van Musschenbroek at Leyden. Thus, at Franeker it was introduced by 's Gravesande's ex-pupil the professor of philosophy Johannes Oosterdijk Schacht in 1727,[51] and at Harderwijk in 1734 by another professor of philosophy A. J. Drijfhout.[52]

   The greatest influence on the growth of experimental philosophy in the Dutch Republic was Robert Boyle. It was his empiricism which caused the evolution of Dutch Cartesian rationalism into the eighteenth-century experimentalism of which the major early exponents were Boerhaave, 's Gravesande and Petrus van Musschenbroek. An added factor to this development was the obvious successes made by the Royal Society using this method. Although Newton's *Principia* and *Opticks* were read in the Republic as soon as they became available by such scholars as Christiaan Huygens, De Volder and the theologian Bernard Nieuwentijt, no references to these works are to be found in the Dutch experimental philosophy courses until 's Gravesande's textbook on Newtonian physics published in two volumes in 1720 and 1721. This work was immediately received with much enthusiasm throughout Europe and a number of foreign translations appeared, including a very popular English version by J. Th. Desaguliers.[53] The main impact of 's Gravesande and Petrus van Musschenbroek was their textbooks interpreting the Newtonian experimental philosophy while they made little theoretical contribution. Their skill was in their teaching and they were to a large measure responsible for the increasing popular interest in science in the Dutch Republic during the second quarter of the eighteenth century. It was with this in mind that Petrus van Musschenbroek enlarged his *Elementa physices* in a Dutch edition especially intended for the layman who could not read Latin.[54]

Both 's Gravesande and Petrus van Musschenbroek had their own instrument collections. Much of their apparatus was made by Jan van Musschenbroek, the elder brother of Petrus. The van Musschenbroeks were a renowned family of instrument-makers in the seventeenth and eighteenth centuries of whom Jan was the best known.[55] Besides being an excellent instrument-maker, he was also a good mathematician and a proficient linguist. On several occasions he was offered professorships in philosophy and mathematics, but he refused to leave his native Leyden. He became a great friend of 's Gravesande, and together they designed many of the instruments described in 's Gravesande's *Physices elementa mathematica*, and used to such good effect in the latter's lecture-demonstrations. In the design of these instruments, they were undoubtedly influenced by the work of the Hauksbees, Whiston and Desaguliers.[56] When 's Gravesande extended his visit to London after the Coronation of George I in 1715, he met Newton and others members of the Royal Society and attended their meetings.[57] It would not be very surprising if during his stay he went to the popular scientific lecture-demonstrations expounding the Newtonian physics, which were given by the younger Hauksbee and Whiston at the former's house in Crane Court near the Royal Society. In the Introduction of his *Physices elementa mathematica*, 's Gravesande acknowledges his debt to the English in his use of the "experimental method": "I shall always glory in treading in their Footsteps, who, with the Prince of Philosophers [Newton] for their Guide, have first opened the Way to the Discovery of Truth in Philosophical Matter."[58]

It is probably true to say that with the able assistance of Jan van Musschenbroek, 's Gravesande was the first to publish a *coherent* lecture-demonstration course in experimental physics. Looking at 's Gravesande and the van Musschenbroek family, one cannot help but be struck by the close connexion in the eighteenth century between the natural philosopher and the instrument-maker. The van Musschenbroeks are a particularly good example, for this one family produced three of the finest Dutch instrument-makers and one of the best eighteenth-century experimental physicists, who, like his predecessor 's Gravesande, fully realised the importance of lecture-demonstrations as a didactic tool. It cannot be stressed often enough that many of the eighteenth-century instrument-makers were not mere *artificers* but were scientists in their own right. They formed a part of the intellectual climate in which experimental physics developed. The fragments of correspondence between Jan van Musschenbroek and his younger brother Petrus are particularly revealing in this respect, dealing on the whole with scientific problems. On several occasions Petrus asks Jan's advice about such questions as the effect of a vacuum on fire,[59] the causes which prevent two thermometers from giving the same readings[60] and about specific problems in mechanics. He is also quite capable of pointing out errors committed by his learned brother in his textbooks.[61]

A number of other examples could be given such as James Short, Jesse Ramsden and Edward Nairne, who specialised in a specific scientific area in which they made significant contributions for which they were elected Fellows of the Royal Society, and who supported themselves by the sale of their instruments. Others such as Benjamin Martin and George Adams (Sr.) wrote textbooks, which, if they did not advance scientific theory, certainly contributed to the popularisation and diffusion of scientific knowledge. In Holland both Fahrenheit and John Cuthbertson made theoretical contributions to their chosen fields, sold scientific instruments and gave public lectures.[62] Furthermore, the advances in instrument design made by these men resulted in the accumulation of new scientific data which could then be synthesised by the theoretician. Thus, the interaction between the natural philosopher and the instrument-maker was a complex one, which in the past tended to be overlooked by those historians of ideas who saw the latter as mere purveyors of apparatus. Like the eighteenth-century natural philosophers, the instrument-makers were not a homogeneous group, but consisted of various types with different talents and ideals.

Jan van Musschenbroek's workshop was internationally famous and was visited by such well-known travellers as von Uffenbach[63] and von Haller.[64] His instruments, identical to those depicted in 's Gravesande's textbooks, were purchased by the Landgrave Carl von Hessen, by the universities of Franeker and Utrecht, and by many private collectors.[65] These instruments were also copied from the detailed engravings in 's Gravesande's *Physices elementa mathematica* by both Dutch and foreign instrument-makers. His price list appended to Petrus van Musschenbroek's *Beginselen der Natuurkunde* contains a great variety of almost two hundred instruments, and it is impossible to determine to what extent he sub-contracted their manufacture. Included are barometers and thermometers, and these were almost certainly made by D. G. Fahrenheit who also manufactured thermometers for 's Gravesande as Jan was no glassblower.[66] On the latter's death, his substantial instrument collection was purchased for 3981 guilders by Leyden University.[67] When Petrus van Musschenbroek died in 1761, he left a very fine scientific library and a large collection of instruments. Both were auctioned in 1762. The auction catalogue of the instruments lists 656 items of physical apparatus, fourteen anatomical and surgical instruments, and twenty chemical items including preparations.[68] This collection was not bought by the university, although a number of the items may have been purchased by them through an agent.

Several other important factors contributed to the accelerating interest in science in the Dutch Republic during the early part of the eighteenth century. These were the popularisation of science, its closely connected sister movement based on the argument that the study of nature demonstrated the power and wisdom of the divine Creator, and related to these the emergence of the Dutch scientific societies. Popular lectures started in England at the turn of the century and in the Republic soon afterwards. Little research has to date been devoted on this topic in the Dutch context. The first known public lecture-demonstrations were given in Amsterdam from 1718 to 1729 by D. G. Fahrenheit,[69] and he was soon followed by others, including Martinus Martens in 1736 and in the same year by the physician Leonard Stocke at Middelburg.[70] The value of these courses varied enormously. Some, like those of Martens, were carefully planned and were the precursors of adult education, while others consisted simply of a few tricks shown at the fair or in the local tavern. In this latter case, the startling experiments shown by the electrical machine became especially popular.

The popularisation of science really started in Holland as a result of a lecture-tour made by Desaguliers in 1729 and 1730.[71] Many of the founders of the early Dutch scientific societies went to his lectures and also members of the universities. Thus, Boerhaave may have listened to him while Petrus van Musschenbroek referred to Desagulier's popularity in the Preface of his *Beginselen der Natuurkunde*. It has been suggested by Nicholas Hans that there exists a connexion between the eighteenth-century movement to popularise science, educational reform and free masonry. According to him, English speculative free masonry was very keen on the diffusion of knowledge, and he suggests that Desaguliers as the English third Grand Master in 1719, was very prominent in this development.[72] This is, however, not really borne out by Gould's history of free masonry.[73] Hans also suggests that a number of the Dutch academics who were interested in educational reform, such as Boerhaave, 's Gravesande and Petrus van Musschenbroek, were probably initiated by Desaguliers into free masonry, but certainly no proof has been found by him or by the present author to substantiate this.[74] Thus, the possible connexion between the early eighteenth-century movement to popularise science and free masonry remains a tenuous one. The first Dutch lodge was established in 1734 and there exists a certain amount of evidence to connect Dutch free masonry with later educational reform, the founding of schools and charitable and welfare institutions started during the last decades of the eighteenth century, especially through the Society of Public Welfare [*Tot Nut van 't Algemeen*]. This society was one of about twenty which were founded by masons, but they do not appear to have established any societies which dealt specifically with the dissemination of science.[75]

In the Dutch Republic as elsewhere, the popularisation of science appears to have had strong religious connexion, especially as it was argued by Robert

Boyle, Sir Thomas Browne, Newton and many others, that research into the working of nature led to a better understanding of its Creator. The major exponent of natural theology in the Republic was Nieuwentijt, whose main work on this topic went into numerous editions and translations. It was translated into English as *The Religious Philosopher* in 1718 by John Chamberlayne.[76] Other popular works in a like vein were written by Martinet[77] and by Uilkens.[78] Several protestant ministers also took a more active part in this movement. Thus, the Remonstrant minister Noozeman was one of the founders of the *Bataafsch Genootschap der Proefondervindelijke Wijsbegeerte,* the principal scientific society in Rotterdam, and five founder members of the scientific society at Middelburg were ministers, including C. H. D. Ballot who became a professor of philosophy.[79] The Mennonite publisher Isaac Tirron of Amsterdam, published many popular scientific works, including the Dutch translations of Desaguliers.[80] The *Vaderlandsche letteroefeningen* and the *Algemeene letterbode,* both journals which published popular scientific articles and reports of the latest experiments, were started by the Mennonite publishing family Loosjes.[81] Furthermore, a Mennonite teacher-training college which taught science to an advanced level, was founded in 1735. It was hoped that this would keep the students training for the ministry abreast with the latest scientific developments.[82]

The emergence of the scientific societies was important as apart from the universities and a few rich individuals, only these could afford to set up laboratories. They resulted in the concentration of groups of natural philosophers, which meant that scientific effort could be coordinated, that the experimental results were generally published, and that financial aid could be given to the individual scientist with no private means.[83] About twenty major scientific societies were founded in the Republic during the second half of the eighteenth century, such as the *Hollandsche Maatschappij der Wetenschappen* at Haarlem in 1752, and similar ones at Middelburg in 1765, at Rotterdam in 1769, at Utrecht in 1773, at Amsterdam in 1777 and 1788, at The Hague in 1793, at Leeuwarden in 1795, at Zutphen in the following year and at Groningen in 1801. The histories of only a few of these societies have been recorded, and as yet no work exists placing all these societies in an historical context.[84] The structure of the *Hollandsche Maatschappij der Wetenschappen* demonstrates how seriously some of the Dutch societies took their task of education and research. Its membership consisted of two groups; a large number of fee-paying Directors who made the policies, and the scientific members who paid no subscriptions but had to contribute regularly to the society's Proceedings if they wanted to remain members, although in practice this rule was never enforced. They could use the society's resources but had no say in policy decisions.[85] Martinus van Marum was elected a member of this society in 1776. In 1784 he was appointed the Director of the fine instrument collection of a second society in Haarlem known as the

Teylers Tweede Genootschap, after its founder, the wealthy merchant Pieter Teyler van der Hulst. In this position van Marum could continue to build up this collection which ultimately was one of the finest experimental physics collections in Europe and devote all his time to research, notably in chemistry and electricity. Men like him can be regarded as the first professional scientists outside the universities although their roles were very similar.[86] Often they had to give public lecture-courses arranged by the society and also advise the municipal authorities about scientific matters. However, by far most of the eighteenth-century men of science were "amateurs" in the sense that they were not employed as scientists, but this did not affect their abilities.[87] Dutch contemporary literature sometimes refers to eighteenth-century science in a rather derogatory way as "salonwetenschap" [drawingroom-science], but it was this general popular interest which ultimately could make the financial contributions necessary for future scientific development.[88]

References

1. This paper can only give an indication of the Dutch material that exists on this topic; some idea of its richness is given by the following foreign works: P. Dibon (1954), *La Philosophie néerlandaise au siècle d'or.* Tome I: *L'Enseignement philosophique dans les universités à l'époque pré-cartésienne, 1575–1650*, Paris, is very thorough and it is a pity that his second volume dealing with Dutch Cartesianism is not yet ready. E. C. Ruestow (1973), *Physics at 17th and 18th-Century Leyden*, The Hague, is again very detailed although it makes little attempt to relate Leyden with science education in other countries, nor does it use archival material but is mainly based on published dissertations (disputations). An interesting work which relates Dutch universities (especially Leyden) with German culture, is H. Schneppen (1960), *Niederländische Universitäten und Deutsches Geitesleben von der Gründung der Universität Leiden bis ins späte 18. Jahrhundert*, Münster. An account of the transmission of Newtonian science to France via Holland is given by P. Brunet (1926), *Les physiciens Hollandais et la méthode expérimentale en France au XVIIIe siècle*, Paris. For a short account of early Dutch influences on English culture see D. W. Davies (1964), *Dutch influences on English culture 1558–1625* (Cornell University Press: The Folger Shakespeare Library), which gives a useful critical bibliography on pp. 35–8. A brief sketch which demonstrates that these influences went in both directions has been given by T. H. Levere (1970), "Relations and rivalry: interactions between Britain and the Netherlands in eighteenth-century science and technology", *History of science*, **ix**, 42–53, with a select bibliography.
2. See for instance J. L. Motley (1906), *The rise of the Dutch Republic*, 3 vols., London and New York, for an early English classic on this topic. P. Geyl (1930/34), *Geschiedenis van de Nederlandsche stam* [to 1688], 2 vols., Amsterdam, and *Dutch civilisation in the seventeenth century, and other essays, selected by P. Geyl and F. W. N. Hugenholtz*, London, 1968. J. Huizinga (1949/53), "How Holland became a nation", in *Verzamelde werken*, 9 vols., Haarlem, **ii**, 266–83, which also contains a number of other useful papers on Holland in a European context, such as

"Duitsland's invloed op de Nederlandse beschaving", in **ii**, 304–49. For a very brief account see C. H. Wilson (1968), *The Dutch Republic*, London, who is better known for his *Anglo-Dutch commerce and finance in the eighteenth century* (Cambridge, 1941).

3. Secondary education was reorganised at a provincial level. Friesland in 1588 produced the first "schoolorder" after this had been suggested by the Prince of Orange, but the most influential one was the one formulated by the university of Leyden for the provinces of Holland and Zeeland and passed in 1625. The main aims were to unify the teaching in all the schools and to produce a single set of textbooks for the protestant children of the new Republic. It had been hoped that the 1625 "schoolorder" would have been passed by the States-General for the *whole* country, but this did not happen. See E. J. Kuiper (1958), *De Hollandse "Schoolorde" van 1625, een studie over het onderwijs op de Latijnse Scholen in Nederland in de 17de en 18de eeuw* (dissertation, Amsterdam: printed by J. B. Wolters of Groningen), 39–57, 78, 146; and H. W. Fortgens (1958), *Schola Latina, uit het verleden van ons voorbereidend hoger onderwijs*, Zwolle, 26–31.

4. A. A. M. de Haan (1960), *Het wijsgerig onderwijs aan het Gymnasium Illustre en de Hogeschool te Harderwijk, 1599–1811* (dissertation, Leyden: printed by "Flevo" v.h. Gebr. Mooij of Harderwijk), 66f.

5. A detailed account of this episode was given by A. de Hoog, research student at Worcester College, Oxford, in an unpublished paper entitled "Cartesianism, politics and the Dutch universities", read on 8 November 1973. The resolution made by the States of Holland on 30 September 1656 was entitled "Ordre jeghens vermenginge van de Theologie met de Philosophie, ende het misbruyck van de vryheyt in het Philosopheren tot nadeel van de Heylige Schriftuyre." [Regulation concerning the mixing of Theology with Philosophy, and the abuse of the freedom of Philosophising to the detriment of Holy Scriptures.] The quotation is a paraphrase of the original text. See P. C. Molhuysen, "Bronnen tot de geschiedenis der Leidsche Universiteit", *Rijks geschiedkundige publicatiën*, **xxxviii** ('s Gravenhage, 1918), 58*, the complete document is reproduced on pp. 47*–58*, and also in Siegenbeek (ref. 7), **ii**, Bijlage VI, 343–62.

6. de Haan, ref. 4, 59f. H. Bouman (1844/7), *Geschiedenis van de voormalige Geldersche Hoogeschool en hare hoogleeraren*, 2 vols., Utrecht, **i**, 330f.

7. M. Siegenbeek (1829/32), *Geschiedenis der Leidsche Hoogeschool, van hare oprigting in den jare 1575, tot het jaar 1825*, 2 vols., Leiden, **i**, 213.

8. G. W. Kernkamp (1933), "Pieter Burman, van 1696 tot 1715 hoogleeraar te Utrecht", *Verslag van het verhandelde in de algemeene vergadering en de sectie-vergaderingen van het Provinciaal Utrecht Genootschap van Kunsten en Wetenschappen*, Utrecht, 86ff. See also Chr. Sepp (1873/4), *Het godgeleerde onderwijs in Nederland gedurende de 16de en 17de eeuw*, 2 vols., Leiden, **ii**, 346ff, 390ff; and J. A. Cramer (1932), *De theologische faculteit te Utrecht ten tijde van Voetius*, Utrecht, 119f.

9. J. G. van Dillen (1954), "Honderd jaar economische ontwikkeling van het Noorden 1648–1748", *Algemeene geschiedenis der Nederlanden*, **vii**, 98f.

10. *The works of Andrew Marvell* (1726) (ed. T. Cooke), London, 124, in a poem entitled "The Character of Holland".

11. H. J. Koenen (1846), *Geschiedenis van de vestiging en den invloed der Fransche vlugtelingen in Nederland*, Leiden, 95–128.

12. See ref. 3.

13. Siegenbeek, ref. 7, **i**, 2ff.

14. For literature on the illustrious schools, see L. R. Hermans (1852), "Geschiedenis der Illustre en Latijnsche Scholen te 's-Hertogenbosch van haar ontstaan in den

jare 1630 tot hare opheffing in den jare 1848", *Bijdr. tot de kennis en den bloei der Ned. Gymn.*, Amsterdam, 55–175, and by F. L. R. Sassen (1963), "Het wijsgerig onderwijs aan de Illustre school te 's-Hertogenbosch, 1636–1810", *Meded., K. Nederlandse akademie van wetenschappen, afdeling letteren*, nieuwe reeks, **xxvi**, Amsterdam; "Het wijsgerig onderwijs aan de Illustre school te Breda, 1646–1669" *as above*, nieuwe reeks, **xxv**, Amsterdam, 1962; "De Illustre school te Maastricht en haar hoogleraren, 1683–1794", *as above*, nieuwe reeks, **xxxv**, Amsterdam, 1972; all these are very well annotated; and his "Levensberichten van de hoogleraren der Kwartierlijke Hogeschool te Nijmegen", *Numaga*, **ix** (1962), 113–115. J. G. van Slee (1916), *De Illustre school te Deventer* ('s-Gravenhage). G. D. J. Schotel (1857), *De Illustre school te Dordrecht*, Utrecht. The Amsterdam "Doorluchtige School" or Athenaeum became the Rijksuniversiteit in 1877, see H. J. van der Beek (1963), *E. H. von Baumhauer zijn betekenis voor de wetenschap en de Nederlandsche economie* (dissertation, Leiden), 12–18.

15. J. Huizinga (1914), "Eerste gedeelte: Geschiedenis der Universiteit gedurende de derde eeuw van haar bestaan", in *Academia Groningana MDCXIV–MCMXIV. Gedenkboek ter gelegenheid van het derde eeuwfeest der Universiteit te Groningen in opdracht van den academischen senaat*, Groningen, 44f.

16. Huizinga, ref. 2, **viii**, 342f.

17. W. J. A. Jonckbloet (1864), *Gedenkboek der hogeschool te Groningen*, Groningen, 56.

18. Ref. 17, 56, 442.

19. Ref. 17, 57, 302.

20. Siegenbeek, ref. 7, **i**, 294; **ii**, 160.

21. F. W. T. Hunter (1927/1943), *Charles de L'Escluse (Carolus Clusius) Nederlandsche Kruidkundige 1526–1609*, 2 vols., 's-Gravenhage, **i**, 187–93.

22. J. E. Kroon (1911), *Bijdragen tot de geschiedenis van het geneeskundig onderwijs aan de Leidsche Universiteit, 1575–1625* (dissertation, Leiden: printed by S. C. van Doesburg), 91–107.

23. E. Peacock (1883), *Index to English speaking students who have graduated at Leyden University*, London.

24. R. W. Innis Smith (1932), *English-speaking students of medicine at the University of Leyden*, Edinburgh and London.

25. D. Guthrie (1959), "The influence of the Leyden school of medicine upon Scottish medicine", *Medical history*, **iii**, 109f.

26. Ref. 25, 115.

27. A. Rook (1969), "Medicine at Cambridge, 1660–1760", *Medical history*, **xiii**, 257.

28. Ref. 27, 260ff.

29. For a full analysis of the places of origin of these foreign students see Jonckbloet, ref. 17, 46–50.

30. H. Bouman (1844/7), *Geschiedenis van de voormalige Geldersche Hogeschool en hare hoogleeraren*, 2 vols., Utrecht, **i**, 186ff, 358ff. For an analysis of the places of origin of the German students see Schneppen, ref. 1, 9–57.

31. H. McLachlan (1931), *English education under the Test Acts*, Manchester, 19f. J. W. Ashley Smith (1954), *The birth of modern education. The contribution of the Dissenting Academies 1660–1800*, London, 64–7, gives a short and not very accurate summary of the influence of the Netherlands on English education and thought.

32. M. Rooseboom (1950), "Leeuwenhoek, the man: a son of his nation and his time", *Bulletin of the British society for the history of science*, **i**, 79–85, and her "Antoni van Leeuwenhoek zijn ontdekkingen en het denken van zijn tijd", *Spiegel historiael*, **iii** (1968), 13–21.

33. Letter of 4 May 1679, in *Oeuvres complètes de Christiaan Huygens* (The Hague,

1899), **viii**, 159, no. 2167 and cited by Rooseboom, ref. 32, 83.

34. P. van der Star (1953), *Descriptive catalogue of the simple microscopes* (Communication no. 87 of the National Museum of the History of Science, Leiden), describes the Leeuwenhoek microscopes and similar ones made by Johan van Musschenbroek which are in the possession of this museum. It also has a number of lenses made by Christiaan Huygens and by his elder brother Constantijn and several clocks made to Christiaan's design, including the oldest known pendulum clock by Salomon Coster, dated 1657 which is the year Christiaan was granted the patent for his pendulum. These items were retained by the Huygens family until 1754 when they were auctioned. See C. A. Crommelin (1949), *Descriptive catalogue of the Huygens collection in the Rijksmuseum voor de Geschiedenis der Natuurwetenschappen* (Communication no. 70, Leiden), 20f.

35. R. Hooykaas (1960/1), "De Baconiaanse traditie in de natuurwetenschap", *Algemeen Nederlands tijdschrift voor de wijsbegeerte en psychologie,* **liii**, 182ff.

36. E. J. Dijksterhuis (1924), *Val en worp* (Groningen), 304–21, and next ref.

37. R. Hooykaas (1950/2), "Science and religion in the seventeenth century (Isaac Beeckman)", *Free university quarterly,* **i**, 169–83.

38. E. J. Dijksterhuis (1970), *Simon Stevin science in the Netherlands around 1600* (The Hague), 126–9, based on Stevin's "Uytspraeck over de Weerdicheyt der Duytsche spraeck" which forms the Introduction of *De Beghinselen der Weeghconst* (Leiden, 1586).

39. C. L. Thijssen-Schoute (1954), "Lodewijk Meyer en diens verhouding tot Descartes en Spinoza", *Meded. van wege het Spinozahuis*, **xi**, 5ff, 25. Meyer wrote the Preface to Spinoza's work on Descartes's philosophy, published in 1663, and the following year this work was translated into Dutch by Pieter Balling. Meyer's ideas about the vernacular are discussed in his "Voorreeden" to *L. Meyers Woordenschat* (Amsterdam, 1669), an extended edition of J. Hofmans *Nederlantsche woordenschat dat is verduytschinge van uytheemsche woorden, die somtijdts onder het Nederlandtsch gevonden worden* (Amsterdam, 1654), many of those dealing with science and technology were taken from Stevin.

40. P. C. Molhuysen (1916), "Over de graden die oudtijds aan de Leidsche Universiteit werden verleend", *Leidsch jaarboekje voor de geschiedenis en oudheidkunde van Leiden en Rijnland,* **xiii**, 31, and Jonckbloet, ref. 17, 278ff.

41. S. H. M. Galama (1954), *Het wijsgerig onderwijs aan de Hoogeschool te Franeker, 1585–1811* (dissertation, Leiden: printed by T. Wever of Franeker), 28.

42. M. Foster (1901), *Lectures on the history of physiology during the sixteenth, seventeenth and eighteenth centuries*, Cambridge, 147. T. Puschmann (1966), *A history of medical education*, New York and London, 368ff. On the origins of the chemical laboratory at Leyden, and for further references of the Dutch sources, see G. A. Lindeboom (1968), *Herman Boerhaave the man and his work*, London, 111–113.

43. Siegenbeek, ref. 7, **i**, 116f.

44. Ruestow, ref. 1, 96f. Molhuysen, ref. 5, **iii**, 298ff.

45. Jonckbloet, ref. 17, 402f. This university may have possessed some instruments as early as 1617, but in 1668 negotiations were started to purchase the instruments of the surveyor Harm Willems.

46. Ref. 45, p. 402, and J. MacLean (1972), "Science and technology at Groningen University (1698–1702)", *Annals of Science,* **xxix**, 187–201, recounts Bernoulli's controversy with the theology faculty by whom he was accused of being a Socinian, but MacLean omits to point out that another reason for these disagreements were Bernoulli's use of the university chapel for his experiments.

47. De Haan, ref. 4, 72.

48. Galama, ref. 41, 139.

49. G. W. Kernkamp (1938), "Acta et decreta senatus vroedschapsresolutiën en andere bescheiden betreffende de Utrechtsche Academie", *Werken uitgegeven door het Historisch Genootschap, Utrecht*, 3rd series, **lxviii**, 207, 595–624.

50. H. J. M. Bos (1967), *Descriptive catalogue. Mechanical instruments in the Utrecht University Museum,* Utrecht.

51. Galama, ref. 41, 153f.

52. De Haan, ref. 4, 101f, and Bouman, ref. 30, **ii**, 298–303.

53. As Christiaan Huygens died in 1695, he could only have read the *Principia*. De Volder referred to this work in his *Oratio de rationes viribus, et usu in scientiis* (1698), but as he retired in 1705 he could not react professionally to the *Opticks*. The "experimental method" of the *Opticks* was much appreciated by natural theologians such as Nieuwentijt. For academic reaction, see W. J. 's Gravesande (1720/1), *Physices elementa mathematica experimentis confirmata size introductio ad philosophiam Newtoniam*, 2 vols., Leiden, and translated by J. Th. Desaguliers (1720/1), *Mathematical elements of natural philosophy, confirm'd by experiments; or an introduction to Sir Isaac Newton's philosophy*, 2 vols., London. I have used the second edition of 1726.

54. P. van Musschenbroek (1734), *Elementa physicae, conscripta in usus academicos* (Lugdum Batavorum), enlarged as *Beginselen der natuurkunde, beschreven ter dienst der Landgenoten...waarbij gevoegd is eene beschrijving der nieuwe en onlangs uytgevonden luchtpompen, met haar gebruyk tot veele proefneemingen door J.V.M.* (Leiden, 1736).

55. C. A. Crommelin (1951), *Descriptive catalogue of the physical instruments of the 18th century, including the collection 's Gravesande-Musschenbroek* (Communication no. 81 of the National Museum of the History of Science, Leiden), 11–15. M. Rooseboom (1950), *Bijdrage tot de geschiedenis der instrumentmakerskunst in de Noordelijke Nederlanden tot omstreeks 1840* (Communication no. 74 of the same museum, Leiden), 102–9. Ms. "Geslacht-Lyst van Van Musschenbroek" by Jan Willem van Musschenbroek (1727–1807) in the library of the National Museum of the History of Science at Leyden.

56. W. Whiston (1714), *A course of mechanical, optical, hydrostatical and pneumatical experiments to be performed by Francis Hauksbee; and the explanatory lectures read by William Whiston* (London), and several later editions, also undated. Both the elder and the younger Hauksbee were involved with these lectures. See also the Preface of H. Ditton (1705), *The general laws of nature and motion, with their application to mechanics* (London), and M. Farrel (1973). "The Life and Work of William Whiston" (unpublished Ph.D. thesis, Manchester–UMIST), Chapter IV, and G. L'E. Turner (1970), "The apparatus of science", *History of science*, **ix**, 133, 138, ref. 19.

57. Royal Society (London) Journal Book (Copy), **xi**, 57.

58. Desagulier's translation, ref. 53, **i**, xviii.

59. Undated letter from Jan to Petrus bound in Ms. "Collegium physicum" in the Petrus van Musschenbroek Handschriften in the Archives of the Universiteits Bibliotheek Leiden, Ms. BPL 240 (9), ff. pp. 126–7.

60. Ref. 59, Ms. BPL 240 (9), f. p. 185, undated letter.

61. Undated letter, Ms.-Arch. 138e-B50, 4 folios, and Arch. 251C, fragment entitled "Responsiones ad Junini (?) objectiones", 4 folios, unsigned and undated, in the Musschenbroek Archives, Leiden National Museum of the History of Science which are presently being examined by this author.

62. C. A. Crommelin (August 1936), "De Hollandsche natuurkunde in de 18de eeuw en de oorsprong der natuurkundige instrumentmakerskunst", *De gids*, 196–211,

and W. D. Hackmann (1973), *John and Jonathan Cuthbertson. The invention of the eighteenth century plate electrical machine* (Communication no. 142 of the National Museum of the History of Science, Leyden).

63. Z. C. von Uffenbach (1754), *Merkwürdige Reisen durch Niedersachsen, Holland und Engelland*, 3 vols., Ulm, **iii**, 430–7.

64. A. von Haller (1883), *Tagebücher seiner Reisen nach Deutschland, Holland und Engelland 1723–1727* (Leipzig), 107.

65. Rooseboom, ref. 55, p. 103, and the many auction catalogues of private collections listed at the back, pp. 142–56. It is at this stage impossible to say which of these instruments were made by Jan van Musschenbroek.

66. Ref. 65, p. 63ff. Crommelin, ref. 55, p. 14. Wildeman (1960), "Een oude catalogus van instrumenten", *Ned. tijdschrift voor de geneeskunde*, **lx**, 1958ff.

67. Molhuysen (1921), ref. 5, **xlviii**, 246, 140*, 141*. The instruments were valued at 3,981.10 guilders by Jan van Musschenbroek. Siegenbeek, ref. 7, **i**, 273; **ii**, 116.

68. *Collectio exquisitissima Instrumentorum in primis ad physicam experimentalem pertinentium quibus dum irvebat usus fuit vir celeberrimus Petrus van Musschenbroek, A.L.M. Medic., Phil. doct. Philosophiae Matheseos in Acad. Lugd. Bat. Professor ord., etc. etc. etc. etc., quorum auctio fiet in Aedibus defuncti per S. et J. Luchtmans ad diem 15 Martii seqq. 1762*. The copy of this catalogue preserved in the Gemeente Archief (Municipal Archives) of Leyden is annotated with the names of the buyers and the prices fetched at the auction. See Gemeente Archief-Leiden Bibliotheek No. 75562. It is bound behind the *Bibliotheca Musschenbroekiana*; the auction catalogue of his fine library.

69. T. Dekker (1955), "De popularisering der natuurwetenschap in Nederland in de achtiende eeuw", *Geloof en wetenschap*, **liii**, 173.

70. D. Schoute (1923), *De geschiedenis van het Natuurkundig Gezelschap te Middelburg* (Middelburg), 1–6. M. Martens (1741), *Beknopte aanspraak van den Heere Martinus Martens, uitgesproken volgens de jaarlijksche gewoonte, op 6 Feb. 1741, waarin de nut en vermakelijkheid der natuurkunde wordt voorgesteld en aangetoond. Uitgegeven door een liefhebber dezer kennise* (Amsterdam). See also G. W. Kernkamp (1910), "Brengt Ferner's dagboek van zijn reis door Nederland", *Bijdragen van het Historisch Genootschap*, **xxxi**, 359ff.

71. There are numerous references to this highly successful lecture tour in the Dutch literature, for example Schoute, ref. 70, p. 6; Dekker, ref. 69, p. 173f, *Gedenkboek Bat. Gen.* (ref. 84), pp. 1f. For an interesting discussion about popularisation of science and natural theology in Holland, see R. Hooykaas (1946), *Rede en ervaring in de natuurwetenschap der XVIII^{de} eeuw, inaugurele rede aan de Vrije Universiteit 1 Feb. 1946* (Amsterdam), 5ff.

72. N. Hans (1951), *New trends in education in the eighteenth century* (London), 137ff.

73. R. F. Gould (1882/7), *The history of free masonry*, 6 vols., London, **iv**, 315.

74. N. Hans (1965), "Holland in the eighteenth century-*Verlichting* (Enlightenment)", *Paedagogica historica*, **v**, 19f.

75. H. Maarschalk (1872), *Geschiedenis van de orde der vrijmetselaren in Nederland* (Breda), 247f. On the attempts to diffuse science to the masses by the society *Tot Nut van 't Algemeen*, see also N. G. van Kampen (1821), *Beknopte geschiedenis der letteren en wetenschappen in de Nederlanden (van de vroegste tijden af, tot op het begin der negentiende eeuw*, 2 vols., 's Gravenhage, **i**, 616ff, but there is no reference here to any possible influence by the freemasons in this movement.

76. B. Nieuwentijt (1715), *Het regt gebruik der wereltbeschouwingen ter overtuiginge van ongodisten en ongelovigen* (Amsterdam), in seven editions (1750); in English as *The religious philosopher, translated from the Low-Dutch by John Chamberlayne. Adorned with*

cuts (London, 1718, 4th ed. 1730); also in French and German. For a discussion on this topic, see Dekker, ref. 69, 173–88, and J. D. Bots (1972), *Tussen Descartes en Darwin. Geloof en natuurwetenschap in de achtiende eeuw in Nederland* (Assen) which contains a well-balanced and lucid account of natural theology and physico-theology in eighteenth century Holland, analysing especially the work of Nieuwentijt, and with a detailed bibliography.

77. J. F. Martinet (1777/9), *Katechismus der natuur*, 4 vols., Amsterdam, and his *Kleine katechismus der natuur voor kinderen* (Amsterdam, 1779). Both works were extraordinarily popular; of the former there appeared six editions, two shorter versions and a German translation, and of the latter seven Dutch editions, three French, one German and twenty-four English editions entitled, *The catechism of nature for the use of children* (London, 1790) to *post* 1850. See Bot, ref. 76, p. 67f.

78. J. A. Uilkens (1799), *De kennis van den Schepper uit Zijne schepselen, of korte schets der natuurkennis voor de jeugd; tot een grondslag van alle godsdienstige onderwijs* (Groningen) and his *De volmaaktheden van den Schepper in Zijne schepselen beschouwd, tot verheerlijking van God en tot bevordering van nuttige natuurkenning. In redevoeringen* (Groningen, 1803), published until 1822. He also published updated versions of Martinet's books listed in the previous ref.

79. Schoute, ref. 70, p. 15ff; Dekker, ref. 69, p. 176f.

80. D. Bierens de Haan (1883), *Bibliographie néerlandaise historique-scientifique des ouvrages importants dont les auteurs sont nés aux 16ᵉ, 17ᵉ et 18ᵉ siècles, sur les sciences mathématiques et physiques avec leurs applications* (Rome; Nieuwkoop photochemical reprint, 1960, 1965), 74, for Dutch translations of Desaguliers.

81. L. Knappert (1912), *Geschiedenis der Nederlandse Hervormde Kerk gedurende de 18ᵉ en 19ᵉ eeuw* (Amsterdam), 133f.

82. S. Muller (1850), *Geschiedenis van het onderwijs in de theologie bij de Nederlandsche Doopsgezinden* (Amsterdam: Jaarboekje Doopsgezinde Gemeenten), 51ff, 65ff, 93.

83. M. Ornstein (1963), *The role of scientific societies in the seventeenth century* (London: reprint of the 3rd ed., Chicago, 1938), 260ff.

84. As yet no really detailed history exists of the emergence and importance of the Dutch scientific societies, although several of the societies have been written, for example Schoute, ref. 70, J. P. Keunen (1919), "Gedenkboek van het Bataafsch Genootschap der proefondervindelijke wijsbegeerte te Rotterdam 1769–1919. Het aandeel van Nederland in de ontwikkeling der natuurkunde gedurende de laatste 150 jaren", *Nieuwe verh. Bat. Gen. 2ᵉ Reeks*, **viii**; C. P. Burger, *Honderd jaar van het Natuurkundig Genootschap te Leeuwarden* (Leeuwarden: Rede uitgesproken in de vergadering van het Genootschap den 10ᵉ December 1895 door den voorzitter); A. Schierbeek (1943), *Grepen uit de geschiedenis van de Natuurkundige Maatschappij "Diligentia" 1793–1943* (Den Haag); B. de la Faille (1876), *Toespraak bij de gelegenheid van het 75-jarige bestaan van het Natuurkundige Genootschap te Groningen* (Groningen). See also R. P. W. Visser (1970), "De Nederlandsche geleerde genootschappen in de 18ᵉ eeuw", *Werkgroep 18ᵉ eeuw, documentatieblad*, **vii**, 7–18.

85. J. D. Bierens de Haan (1952), *De Hollandse Maatschappij der Wetenschappen 1752–1952* (Haarlem).

86. For Van Marum's chemical research, see T. H. Levere (1969), "Martinus van Marum and the introduction of Lavoisier's chemistry into the Netherlands", in *Martinus van Marum: life and work* (ed. R. J. Forbes) (Haarlem), **i**, 158–286, and his electrical experiments, W. D. Hackmann (1971), "Electrical researches", in the same series, **iii** (Haarlem), 329–78. For his large collection of scientific instruments, see G. L'E. Turner (1973), *Descriptive catalogue of Van Marum's scientific instruments in Teyler's Museum* (Haarlem) which is also published as Part II in **iv** of the series

(Haarlem). Other aspects of Van Marum's scientific work are also discussed in these volumes and elsewhere.

87. This point is now well-recognised by most historians of science, see for instance, D. S. L. Cardwell (1972), *The organisation of science in England* (London: Open University Set Book), 17.

88. R. J. Forbes, "Science in Van Marum's world", in Forbes, ref. 86, **i**, pp. 127–57, especially p. 150, raises a number of points referred to in this paper in the context of Van Marum and the eighteenth century, and Hooykaas, ref. 71, p. 6ff.

7. The Rise and Fall of Scottish Science

J. R. R. CHRISTIE
(University of Leeds)

Since the publication of George Davie's *The Democratic Intellect*, a number of stimulating studies in the intellectual and social history of the Scottish Enlightenment have appeared. Of particular importance are articles by Geoffrey Cantor, Richard Olson, Jack Morrell, Steven Shapin and Nicholas Phillipson.[1] Historiographically the interesting thing about these studies is their refusal to fall into any neat classification along "internal" and "external" lines, a direct consequence, I think, of the manifest social role and status which knowledge-seeking achieved in eighteenth-century Scotland. By deliberately recognising and centralising the social role of knowledge, this essay seeks by implication to raise further doubts about the validity of the internal–external demarcation for a believable historiography of science. More substantially, by attempting to draw together the conclusions of the above scholars without, I hope, doing them too much violence, it tries to imagine what an integrated overview of the history of Scottish science may one day look like. This is doubtlessly an incautiously premature and partial exercise; its value lies in any interpretive and methodological challenge it may have for scholars whose further research will surely test and modify its conclusions.

> "Dr. Black is deprived of his grand discovery of latent heat. . . . The names of our two Hunters, and our two Monros, are just mentioned in half a line. . . . The founders and illustrators of the Huttonian theory are treated with the most marked injustice . . . [they] are blotted out of history by the lethal stroke of Mr. Whewell . . . the fathers of Scottish science have thus fallen in their own field of glory. . . . With such men [as Whewell] we have no community of feeling. They exist chiefly in the cloisters of antiquated institutions, whose prejudices even a pure religion has not been able to abate, and through whose iron bars the light of knowledge and of liberty has not been able to penetrate."[2]

When starting to consider the rise of Scottish science, I thought light might be thrown on the topic by also considering Scottish science in a declinist context. David Brewster's review of Whewell's *History of the Inductive Sciences* suggests that all was not well with the Scots in the 1830s, an impression confirmed by other scientific reviewers; for example, the way in which an approving review of Lyell's *Elements of Geology* modulated into a vehement plea for Hutton's claims as geology's first true theorist.[3] This seeming narrow and bellicose scientific nationalism of the 1830s also reminded me of something else—precisely what I was unable then to put my finger on, but which will be elucidated by the end of this paper. Viewing Scottish science in its declinist phase did at least reveal an anomaly in my whole approach to Scottish science. While the rise of Scottish science was being conceived of in vaguely quantitative terms, the declinist phase was being thought of not in terms of quantity (if anything there was more scientific endeavour in Scotland after 1800 than before), but in terms of the erosion of a philosophical and educational programme. The quantitative approach to the rise of Scottish science is justifiable as providing a partial index for a "take-off" period (1707–40), but once the activities of the Physicians' College, the Surgeons' Incorporation and the Aristotelian, Cartesian and Newtonian teachers of the later seventeenth century are taken into account,[4] purely quantitative claims which see a lot more science in 1740 than in 1690 will have to be heavily qualified. Thus in order to harmonise my views of the rise and decline of Scottish science, quantitative parameters have been abandoned to discuss Scottish science instead in terms of a distinctive philosophical and educational programme which had its roots in the fundamental preoccupations of Scottish lowland society, and which owed its rise, its decline and its distinctiveness to the nature and force of those preoccupations.

Scottish science rose to prominence in the eighteenth century along with numerous other enterprises in literature, history, epistemology, moral and social thought, a collectivity generally referred to as the Scottish Enlightenment. It did so in an organised and differentiated context, that is, in an institutional infrastructure of reformed universities and learned societies, and in an intellectual context whose roots are easily locatable in late seventeenth century and early eighteenth century England, Ireland and Holland: the science of Newton and Boerhaave, the philosophy of Locke, Berekeley and Hutcheson, the natural law tradition of Grotius and Pufendorf. These institutional and intellectual materials from which the Scottish Enlightenment took off have long been recognised.[5] But only in a very limited sense does a list of materials provide an explanation as to why this striking cultural efflorescence should have occurred. Such a list tells us about origins rather than causes, and so characterises rather than explains.

Dugald Stewart's comment about the Scottish Enlightenment owing its existence to the "constant influx of information and liberality from abroad"[6] would obviously come under the heading of characterisation. Stewart's point has, however, been given explanatory purchase by Hugh Trevor-Roper,[7] who has allied this "influx" to two particular Scottish groups, both, significantly in his eyes, minority groups unrepresentative of the mainstreams of Scottish political and religious life. These groups were the liberal Calvinist clergy who returned with King William at the Glorious Revolution to work as intellectual leaven in the Presbyterian lump, and the Jacobite-Episcopalians who had considerable contact with England and Europe. Free from the narrow nationalism and religious bigotry of their time, they are seen by Trevor-Roper as having been able to receive and spread the new learning, thus triggering a new awareness among the Scots about their nation's situation in time and space, which when allied to the contemporary insights offered by a newly expanding economy, produced the historical sociology of Smith, Ferguson and Millar which Trevor-Roper takes to be the main signature of the Scottish Enlightenment.

This account has much to recommend it, particularly its focus on the strong Jacobite-Episcopalian element in Scottish culture before and after the Union of 1707. But it also faces several difficulties. It helps us to recognise a Pitcairne or a Gregory, but what about a Monro or a Maclaurin? It helps us to see how intellectual seed might have been imported and sown on small patches of fertile ground, but not to understand how the culture of political and religious "outs" became the establishment norm by 1750. And it does not look hard enough at what happened to the Jacobite-Episcopalian programme. In pre-Union times this suffered severe set-backs, its adherents either becoming isolated within Scotland,[8] or forced to emigrate to England,[9] where they joined with like-minded men to form an important element in the Newtonian programme of the late seventeenth and early eighteenth centuries.

In post-Union times the Jacobites suffered political and military disaster, and their cultural norms—the classicism of Ruddiman, their iatro-mechanism of Pitcairne—did not survive after the deaths of the individual proponents. These points suggest that we ought to look elsewhere for the social dynamic of a national cultural movement which from the Union onward drew alike on Whig and Jacobite, Presbyterian and Episcopalian, and only began to assume major cultural proportions when the immediate post-Union Jacobite-Episcopalian resurgence was well on the wane.

Nicholas Phillipson suggests that it is in the parliamentary Union of 1707, its causes and consequences, that we may seek an adequate explanation for the Scottish Enlightenment.[10] The decline of the seventeenth century Presbyterian hegemony and its theocratic pretensions after Cromwell and the Restoration enabled the socially élite classes of Scotland, aristocracy and gentry, to reassert their position as social and political leaders of the country, and in addition, it allowed them to start thinking in radically different terms about Scotland's future. The terms in which that future was eventually defined proved to be material rather than spiritual. Scotland was perceived to be a poor and backward nation when compared with England or Holland. The policy to remedy this situation was one of state patronage to encourage Scottish manufactures, complemented by attempts to open up colonial markets along English and Dutch lines. This policy's bias was revealing.

Economic development was sought, development which would preserve and foster the nation's independence. The Scots wished to avoid a debilitating dependence on foreign imports, and so sponsored native manufactures; and if England would not relax the Trade and Navigation Acts for Scotland, why then Scotland would create her own colonies. It was the failure of all these efforts, pointed up by the famines of the 1690s and the disaster in Darien, which ultimately persuaded the Scottish élite that economic development was simply unattainable within the existing framework of relations with England. Exclusion from, even persecution by the Atlantic trading community and protective tariffs within the British Isles, were a decisive bar to economic progress. The solution was political; closer union with England, with terms admitting Scotland to that privileged economic sector. To gain this, even the preferred solution of political federation was abandoned, to the extent that the Scots sacrificed the nominal institutional guarantee of their political independence by submitting to an incorporating Union of Parliaments.

The pre-Union Scottish parliamentarians were fully aware of the possible consequences of this drastic measure, foreseeing a flood of Scots men and money to metropolitan London, foreseeing Scotland's future tied to the English court and ministry. These fears were indeed partially justified. In addition there were other unpalatable consequences, for the economy did not immediately respond in the expected fashion; in fact sectors of it declined even further under English competition. The Union also left a substantial proportion of the Scottish élite in a state of social and political disorientation, their traditional function undercut by the usurpation of their primary institution, the Scottish parliament. Parallel signs of cultural confusion were also visible, such as the schizophrenic oscillation between Scottish vernacular literacy and patriotic history on the one hand, and the attractions of English Augustanism on the other.[11] It was in the response to this national trauma that the ideological and institutional forms of the Scottish Enlightenment were forged. Gentry and aristrocracy responded by reactivating that programme of economic improvement which had united their predecessors. They did so initially by concentrating on agricultural improvement, moving from there to encourage Scottish manufactures; their improving schemes were pursued collectively in the context of two institutions, a large Society for Agricultural Improvement, and the Board of Trustees for the Encouragement of Fisheries, Arts and Manufactures. This improving ideology was not confined to gentry and aristocracy, but found an even readier adherent in city and educational administrative circles, those sectors which had no landed basis to cushion the immediate impact of the post-Union economic recession.[12] The response here was to remodel and expand university education in order to develop its economic potential by attracting the sons of landed society and foreign students, and preventing the flow of young Scots and their money abroad for educational purposes. It was in this period of post-Union reform, and for these sorts of reasons, that the first generation of Enlightenment scientists were brought to the University of Edinburgh, men like Monro *primus* and Maclaurin who provided courses in modern science with a stress on the uses and morally uplifting nature of scientific knowledge.[13] Other cultural responses were varied, but two strong programmes of lasting importance emerged, one concerned with polite literature, the other, fostered by the Rankenian Society and feeding into the universities from there, concerned with nothing less than the foundations of knowledge.[14] These cultural responses were also improver orientated, deliberate and conscious attempts to improve the individual participants in social, moral and intellectual terms, tightly organised in small institutional enclaves. They were not definable in terms of any particular brand of politics or religion, and socially they drew from a range between the petty-bourgeois and the learned upper-middle class professional. The Scottish Enlightenment as it emerged in the mid-eighteenth century was the result of the fusion of the two foregoing sectors, the aristocratic material improvers and the intellectual improvers. The basis of their fusion was obviously the common ideological link; indeed the perceived connexion between intellectual and material improvement was early on a specific postulate of university reform and appointments policy.[15] Other factors strongly supplemented the fusion process, for example the social links between a learned profession like the law, increasingly involved in cultural activity, and the landed élite.[16] Most important of all was the way in which landed society began to identify with the new intellectual culture, seeing there a source of practical aid, instructive entertainment and modern polite values; and the way in which the intellectuals reciprocally began to articulate and practise material improvement. This process of identification started informally in the 1720s, and by the 1740s and 50s had become fully and concretely institutionalised in societies like the Select and the Philosophical Societies of Edinburgh. The result was that by the 1750s something extraordinary had happened to the role which generalist improving culture was playing in Scottish life. Cultural activity had become a substitute for collective political activity by monopolising its driving ideology and by substituting for its institutional forms. As the articulate custodians of improving culture, the intellectuals had inherited the leading social role and function of the leaders of the pre-Union élite and the post-Union agriculturalists. Their ideals and ideas were becoming increasingly important to the way in which Scottish élite society thought, wrote about and acted for its country; their institutions structured the social lives and ambitions of the social élite. The sharp perception of David Hume was well aware of the Scottish process. As the young lawyers and aristocrats clamoured for admittance to the Select Society, he wrote with a euphoria only just restrained by his habitual irony: "It has become a national concern . . . we are as much solicited by candidates as if we were to choose a member of parliament".[17] Hume's simile hit the nail on the head.

The social roots of Scottish science may be firmly located in the ideology and institutions of Scotland's improving tradition. I want therefore to examine the implications of this central point by showing some important ways in which Scottish science continued to respond to this tradition. But before this can be done it is first necessary to extend and refine our understanding of improvement. So far we can understand it in two related senses: firstly in a civic sense, the attempt by individuals acting in groups to better themselves and their nation, to make of themselves virtuous, knowledgeable, socially concerned citizens collectively pursuing the patriotic goal of modernisation, and secondly, within this civic mode, was the intense preoccupation not merely with immediately useful practical knowledge, but with the uses of various forms of knowledge (metaphysical, moral, epistemological, scientific) in relation to the pursuit of the foregoing individual and collective goals. Improvement so defined is generally taken to be a unified ideological constellation, a broad progressive basis for national advance. But it can be shown how improvement may be held to be not all that broad, and that its unity was always problematical; that from post-Union times it had built into it a set of tensions and a particularised set of aims which when taken into account further our understanding of a scientific enterprise erected on improving foundations.

Through improvement the gentry and aristocracy redefined their collective self-image and reaffirmed their sense of social status and control. But control of what? Obviously their traditional local areas of influence in the agricultural hinterland, but more importantly, they saw themselves as responsible for the control of the nation's economic life.[18] The point becomes crucial when we remember how the pre-Union élite had thought of Scotland's independence in economic terms. To secure the economy was to secure the nation, and for that security parliament was sacrificed. But only, it should be noted again, with strong misgivings. Thus although the Union was seen as the precondition of economic progress the doubts about sacrificing the political superstructure remained to trouble the landed classes, reinforced by English interference after the Union and on occasion throughout the eighteenth century. For the Scottish élite it was therefore of paramount importance to reassert themselves in terms which simultaneously vindicated their self-acknowledged role as progressive social leaders working to create a developing economy within an assimilatory political framework defined by the incorporating Union, and as patriotic guardians of a nation which still possessed strongly independent sentiments as to the worth of its own distinctive modes of religion, law, education and culture. The assimilatory and independence sectors of improving ideology were not necessarily in conflict. Indeed, as Dr Phillipson observes, "for much of the eighteenth century assimilation was regarded not so much as a threat to Scottish life as a stimulus to it".[19] Nonetheless, an ideology which could equate political assimilation, economic development and national independence depended on a high degree of social and political stability to maintain its equilibrium. Conditions could arise which would release the potential of its inherent assimilation—independence tensions.

As an ideological construct, improvement partook strongly of neo-Harringtonian country ideology.[20] This emphasised the role of the independent freeholding gentry as guardians of the nation's integrity, who opposed the corrupting effects of a commercial economy which threatened the influence of a traditionally landed polity; opposed the corruptive centralising and bureaucratising tendencies of government ministry, Court, monarchy and military; and who championed the worth of the landed squirearchy exercising benevolently paternalistic authority over tenantry and local legal and ecclesiastical life. Scotsmen such as Andrew Fletcher of Saltoun, John Clerk of Penicuik and David Hume evinced hostility towards narrow, quick-profit mercantile interests, joint-stock companies, centralised government, standing armies and the growth of the Civil List—the traditional canon of Harringtonian deprecation.[21] The appeal of Harringtonian sentiments to a landed class which had been deprived of its primary institutional form of political expression is easily imagined. In the Scottish context, however, this ideology was not confined to its traditional function of political opposition. It was also, and more importantly, the norm of a provincial élite which had adopted improvement as its active principle. The improvers' economic action was largely defined by their Harringtonian self-image. Stemming from a generalised commitment to develop the economy, that action took on specific forms related to landed Harringtonian interests rather than to visionary commercial and financial enterprises of the Darien variety which had attracted an earlier generation. The most obvious of these specific forms was agriculture. Other typical schemes tended to follow the diversification of landed resources: wool, coal-mining, quarrying, the linen industry and its bleaching subsidiary. In addition, there was the development of planned villages on estates, organised round a manufacturing industry. These enabled the landed classes to increase their range of productivity, and also to create a local market for increased productivity in the agricultural sector.[22] Harringtonian economic improvement may be summarised as being directed towards the development of native resources and skills within the traditional structure of local social control by the landed classes.

If the view of Scottish intellectuals as custodians and articulators of improvement is correct, and if this extended analysis of improvement is correct, there are important implications for our historical understanding of Scottish science. Firstly, we would not merely expect to see Scottish science as owing its social existence to improvement—this much is obvious—but as peculiarly responsive to improved preoccupations, these being interest in the foundations and uses of knowledge, and improvements' particularised economic aims. Secondly, we would expect that scientific culture could become subject to the stresses inherent in improvement ideology, being bound up with notions of, on the one hand, national integrity and independence, and on the other, vulnerability to assimilatory pressure.

The areas with which I will illustrate this thesis are philosophy, mathematics, chemistry and methodology. But first a word of clarification about my analysis of these topics is necessary. It has been established that Scottish intellectuals and intellectual enterprise attained an abnormally high social status and an abnormally important social function in eighteenth century Scotland. This meant that because knowledge and its uses were taken so seriously, knowledge and the activity of knowledge-seeking had an unavoidable ideological significance. Intellectual issues were not merely matters of mildly approving or worried interest, but affairs of considerable and immediate social moment, involving questions seen as fundamental to the proper development of the individual and society. Therefore I see strict attention to the social role of knowledge and of knowledge-seeking as the primary analytic necessity for the history of Scottish science.

Let us begin with the Humean theory of knowledge. Hume's philosophical enterprise drew on a variety of recent and contemporary philosophical discussions, its particular Scottish roots being the Rankenian Society's interest in Berkeley.[23] Hume's contemporaries, particularly the Common Sense philosophers, thought his theory subjectivist, very much in the Berkeleyan tradition. Their interpretation of Hume, which is often difficult to grasp from a modern philosophical standpoint, saw him less in the light of rational and logical analysis than in terms of judgements about the implications of his doctrines[24]— the possible uses of his knowledge. Hume's theory of knowledge could be seen as dangerous, striking at the orthodox roots of individual and social behaviour by advocating scepticism, atheism and necessitarianism. If we narrow the focus of discussion to specific philosophical issues by examining Hume's doctrine of causation, several occasions emerge where Humean causation had important institutional implications; in the Philosophical Society in 1754, when the Newtonian natural philosopher John Stewart came near to accusing Hume in print of introducing his doctrine into the Society, subverting Newtonian orthodoxy and originating the metaphysical crisis which produced Lord Kames' anti-Newtonian theory of motion;[25] secondly in the late 1790s when Henry Brougham's Academy of Physics had to ban discussion of necessary connexion before its members could begin to talk productively about science;[26] and most notoriously in 1805, when John Leslie's election to the Edinburgh Chair of Mathematics was opposed publicly not on the perfectly sensible grounds of his Whiggery and atheism, but on the pretext of his philosophical approval of Humean causation.[27] This last was a very striking example of the manner in which the ideological purchase of an intellectual issue was seen to possess more public leverage than a forthright politico-religious confrontation could provide.

The intense and persistent Scottish attention to the foundations and uses of knowledge initiated by the Rankenians and centralised by Hume and Reid was the philosophical hall-mark of Scottish science. In the development of mathematics this philosophical bias had an important determining effect, firstly through the efforts of Maclaurin to meet Berkeley's arguments about the ungodliness of mathematicians and the shaky foundations of infinitesimal calculus,[28] and secondly through the Common Sense grounding of mathematics in experience.[29] The central philosophical issue was the legitimacy of abstractive processes of reasoning. In France these could be taken confidently for granted by a rational mechanician like D'Alembert,[30] but the Scots had to reassure themselves by reflective philosophical analysis. This resulted in a fair measure of agreement on the empirical warrant for the abstract demonstrative reasoning of mathematics and the certainty which it bestowed. However, the reflective empirical bias of the Scottish discussion, linked significantly by Richard Olson to the Common Sense tradition of moral pedagogy, gave to geometrical methods a philosophical legitimacy which did not apply to algebra. The philosophical and pedagogic preferences of Scottish mathematicians go some way to explain why Scottish mathematics did not develop its own form of analytical algebra, and indeed felt philosophically obligated to reject the heuristic benefits of Lagrangian analysis.[31]

Scottish chemistry was another interesting case of scientific development from specific ideological roots. Institutionally, chemistry came into the reformed University of Edinburgh as a pharmacological appendage to the Boerhaavian professors' medical curriculum. Its raison d'être was to provide a thorough education in therapeutic pharmacy for putative physicians, and as such it partook of the considerable success which the Edinburgh Medical School had achieved by mid-century. Along the way professor of chemistry Andrew Plummer had offered his services as analyst to the Philosophical Society's project for a survey of Scotland's mineral resources.[32] But in Plummer's time, improving chemistry of this nature did not infiltrate into the University, where pharmacy remained entrenched. When William Cullen became a candidate for Plummer's professorship in 1755, he did so with what was, in Scottish terms, a radical new programme for academic chemistry, which endeavoured to free the subject from the inhibitions of medical chemistry and the more extreme reductionist pretensions of Newtonian natural philosophy.[33] Cullen met with considerable hostility from the Edinburgh professoriate, because he was known to oppose not only medical chemistry, but the Boerhaavian orthodoxy of the Medical School,[34] and so could be perceived to threaten the whole intellectual and academic structure upon which Edinburgh's medical success was built. Even after ten years on the Faculty at Edinburgh, Cullen was still being requested by the Town Council to restrain his critical attitude towards Boerhaave, because the reputation and drawing power of the Medical School was so closely associated with Boerhaave's name.[35] Why then was Cullen elected in 1755? Cullen seems to have been highly aware of, and sympathetic towards, the ideological bias of Edinburgh culture in the 1740s and 50s. His support of Hume's candidacy for the Chair of Logic at Glasgow,[36] his work with Lord Kames on scientific agriculture,[37] and for the Board of Trustees on bleaching,[38] helped to gain him the support of élite circles in Edinburgh, but his real political muscle lay with Scotland's leading political power, Archibald Campbell, 3rd Duke of Argyll, with whom Cullen shared an interest in salt purification.[39] Worthy of notice are the specific economic areas to which Cullen's chemical projects related: agriculture, bleaching for the linen industry, salt for fisheries— precisely those areas of improvement pushed most strongly by landed society and the Board of Trustees, the particularised elements of Harringtonian economics. Cullen's acute participation in this ideological tradition, and the nature of the support it gained him, eventually proved stronger than the urban-commercial claims of Boerhaavian orthodoxy and medical chemistry.

By the 1760s Scottish chemistry was becoming known particularly through the work of one man, Joseph Black. Much of the history of Scottish science between 1760 and 1800 can be written in terms of the research traditions which Black initiated in pneumatic chemistry and heat.[40] The extent to which Scottish science defined its own particular subjects and problems, became increasingly a self-referential process, is one good justification for claims about a native and distinctive Scottish science. Indeed, the origins of Black's work on causticity and pneumatic chemistry illustrate such claims, and demonstrate once again the ideological basis of Scottish science. One well-known origin to which Black himself referred was the medico-chemical limewater debate which took place in the Edinburgh Medical School in the 1740s and 50s.[41] But the substance lime was not only of medical interest. It also had considerable economic implications in relation to its uses as an agricultural fertiliser and as a bleaching agent for the linen industry. This industry was a particular favourite of the Board of Trustees, and it faced technical problems stemming from the inadequacy of traditional whitening methods. Francis Home was one mid-century chemist who turned his attention to the problems of bleaching.[42] Cullen was another who, in the early 50s from a technical viewpoint and in the 60s from an economic, engaged this topic. We know that Cullen's early work on bleaching was contemporaneous with Black's early work on causticity, and we know too that Cullen's scientific agriculture involved the investigation of lime.[43] On such evidence, and taking into account the close scientific relations between Cullen and Black at that time, it would seem unreasonable not to admit the contemporary emphasis of improving ideology upon agriculture and bleaching as a further source for Black's work on causticity. This approach would also make sense of the way Black later applied his knowledge of causticity to the technical problems of bleaching.[44]

Historians and philosophers of science have recently emphasised the methodological commitments of Scottish science, rightly stressing the severe

anti-hypothetical bias of Thomas Reid's formulation of Newtonian induct-
ivism.[45] This bias had important Scottish precursors, particularly among the
early medical professors, whose Medical Society reserved the right to withhold
publication of essays which were "deficient in facts".[46] The origins of the
Medical Society and its creative preoccupations were very much bound up
with the pedagogic concerns of the medical professors.[47] Their methodological
strictures were those of teachers whose primary aim was to convey practical,
useful knowledge in the training of men who were to apply that knowledge in
situations which demanded practical cures rather than theoretical elucida-
tion. In the period between the early medical professors and the Common
Sense philosophers, however, Scottish methodological thought placed far more
emphasis on theoretical and speculative elements, an emphasis which we can
trace out in Hume's preface to the Philosophical Society's *Essays and Observations*, in Kames' essay *Of the laws of Motion* and most strikingly of all in Adam
Smith's philosophy of science.[48] Smith saw scientific thought as a predominantly speculative enterprise. Its origin lay in the universal human need to
formulate imaginative solutions, whose function was to allay the psychological
disturbances produced by anomalous observations which interrupted the
smooth and regular flow of mental associative processes. When Reid turned
his attention to the problem of the mental basis of science, he specifically
opposed the "happy conjecture" view of science.[49] Instead, he defined science
as an active, although extremely cautious process of multiple observation and
inductive generalisation, which could rely on the promptings of innate faculties
as to the realist ontological status of matter and power.[50] This was a normative
answer to the descriptive and deterministic associationism of Hume and Smith,
who saw speculative thought, particularly of a causal nature, as a psychological necessity grounded in the constitution of the human mind. Reid saw
speculative thought as detrimental to science. He described it as "building
castles in the air" and as such no proper basis for the progress of knowledge.[51]
This is less the attitude of an uninvolved philosopher than of the pedagogue
responsible for educating a social élite, who wished to be seen as progressive
and practical seekers of knowledge, but who also, according to Reid's critique
of Hume and Smith, needed firm guidance towards the proper means of
obtaining true knowledge, and reassurance as to its practical validity.

    Reid's anti-hypothetical bias emerged as a central buttress in the method-
ological defences of those Scottish scientists who evinced doubts about the
new scientific theories produced in late eighteenth- and early nineteenth-
century England and France.[52] Long held theoretical norms of the Scottish
scientific community were directly challenged by the work of Young and
Fresnel in optics, Lavoisier and his school in chemistry, Lagrange, Laplace
and later on the Cambridge Analysts in mathematical physics. Nevertheless,
an embattled scientific posture organised around a commitment to theoretical
norms which only considerable passage of time revealed to be degenerating can

only be justification for a declinist view of Scottish science after 1800 in an extremely partial and locally temporalised sense. It is not wholly realistic to speak of a nation's science as declining because some of its scientists held some theories which lost their viability. However, the impact of foreign science in Scotland can convert into evidence for a declinist view when further factors are taken into account. These factors were those wider philosophical educational and social values which designed the Scottish response in such a way that certain elements of foreign science were perceived to challenge those values. The crucial point is that the challenge and response occurred precisely at the time when those values themselves were beginning to be eroded by a process of socio-economic and political change which affected the structure of Scottish society and its institutional arrangements.

While economic progress seemed to remain in the Harringtonian mould, and while overt assimilationist pressures appeared minimal, improving ideology could maintain stability. But from the 1770s onward, conditions arose which exposed improvement's inherent assimilationist-independence tensions and surfaced doubts about the social implications of emergent industrial-capitalist society. Potential pressures against independence were realised by an expanding economy, as in the case of the Scottish Excise's assimilation to the machinery of English administration. On the legal front too there was continual assimilationist pressure after 1800.[53] Politically the Scots were becoming increasingly involved in the party political norms of late eighteenth-century England; indeed, even matters of scientific organisation were now firmly embedded in party politics, as recent studies of the origins of the Royal Society of Edinburgh and the Academy of Physics have shown.[54] Adam Smith's embryonic recognition of the alienating effects of specialised division of labour in manufacturing industry, and the similar tendencies in education which he saw as a natural consequence of the norms of commerce, were only the earliest expressions of doubt about the effects of economic development upon society.[55] To state the matter as simply and briefly as possible, that degree of political assimilation seen as a precondition for economic progress had set in motion a juggernaut whose effects were not always palatable nor easy to control. Neither was there in every case a manifest desire to control them. While the gentry resisted pressure to reform the Court of Session Bench, it is equally clear that by the 1780s, having attained their economic millenium, they were not about to organise the nation in any form of resistance which might threaten that attainment. Thus it is not surprising that the ideological mantle of nineteenth-century Scottish culture fell not upon the élite aristocrats and gentry of the 1740s and 50s, but upon the high-bourgeois lawyers and educators, whose careers and ambitions, unlike those of the contemporary aristocracy, remained firmly based in Scotland. These were the men who, while recognising the uncontrollable centralising tendencies of metropolitan London science, and morosely vindicating the real if humbler roles which local institutions like the

Royal Society of Edinburgh could still fulfil,[56] were nonetheless able to organise effective resistance against pressures for the assimilation of their institutional and intellectual life to potentially corrupting English standards.[57] They resisted in precisely the terms we would by now expect them to resist: the foundations and uses of knowledge, as discussed by the philosophers of the high Enlightenment and their successors. The process can be seen quite clearly in general cultural terms: the continuing Scottish insistence on the virtues of a generalist liberal education with a philosophical basis, as opposed to the specialism advocated by both radical English utilitarians and Anglican Tories. And we can see the process operating in a subject as specific as mathematics, where the algebraic analysis of Paris and Cambridge was seen as a mere "mechanical knack",[58] proceeding in a mental "miasma",[59] with no true self-reflective philosophical knowledge involved—a neat extension of Smithian alienation into the field of science. Traditional organisational patterns of thought held as fast as educational and intellectual ones. Brewster's remedy for the decline of *English* science was that typical Scottish Enlightenment structure, "an association of our nobility, clergy, gentry and philosophers".[60] Mention of Brewster returns me to my opening query. What did his tirade against Whewell remind me of? It falls into its ideological slot as just one more of those declinist laments, often cast in militaristic or chivalric terms, uttered by Scots from the Union onward about the injustices visited upon them by their powerful assimilating neighbour; from Belhaven's anti-Union speech, through the Judges' Bill agitation, down to Walter Scott and Brewster himself—". . . the fathers of Scottish science have thus fallen in their own field of glory". "The flowers of the forest are all fled away."[61]

References

1. G. Davie (1961), *The democratic intellect: Scotland and her universities in the nineteenth century* (Edinburgh). G. Cantor (1971), "Henry Brougham and the Scottish methodological tradition", *Studies in the history and philosophy of science*, **ii**, 69–89, and "The Academy of Physics at Edinburgh", *Science Studies*, V (1975 forthcoming). R. Olson (1971), "Scottish philosophy and mathematics", *Journal of the History of Ideas*, **xxxii**, 29–44. J. Morrell (1970), "The University of Edinburgh in the late eighteenth century: its scientific eminence and academic structure", *Isis*, **lxii**, 158–71, and "Professors Robison and Playfair, and the *Theophobia Gallica*; natural philosophy, religion and politics in Edinburgh, 1789–1815", *Notes and records of the Royal Society of London*, **xxvi** (1971), 43–63. S. Shapin (1974), "Property, patronage and the politics of science: the founding of the Royal Society of Edinburgh", *British Journal for the History of Science*, **vii**, 1–41, and "The audience for science in eighteenth century Edinburgh", *History of Science*, **xii** (1974), 95–121. N. Phillipson, "Culture and Society in the eighteenth century province: the case of Scotland and the Scottish Enlightenment", in *The University in Society: Studies in the History of Higher Education* (ed. L. Stone), Princeton, 1974. See also J. Christie

(1974), "The origins and development of the Scottish scientific community, 1680–1760", *History of Science*, **xii**, 122–41.

2. D. Brewster (1837), "Review of William Whewell's *History of the Inductive Sciences*", in *Edinburgh Review*, **lxvi**, 148–51.

3. W. Fitton (1839), "Review of Charles Lyell's *Elements of Geology*", in *Edinburgh Review*, **lxix**, 441–66.

4. A start on this area has been made by Fr. J. Russell (1974), "Cosmological teaching in the seventeenth century Scottish universities", *Journal for the History of Astronomy*, **v**, 122–32.

5. By, for example, J. Clive (1970). See his "The social background of the Scottish Renaissance", in *Scotland in the age of improvement* (eds. N. T. Phillipson and R. Mitchison, Edinburgh).

6. D. Stewart (1854–60), *Dissertation exhibiting the progress of metaphysical, ethical and political philosophy*, in *Collected Works of Dugald Stewart* (ed. Sir William Hamilton, Edinburgh), **i**, 551.

7. H. Trevor-Roper (1967), "The Scottish Enlightenment", *Studies on Voltaire and the Eighteenth century*, **lviii**, 1635–58.

8. A. Pitcairne, who was expelled from the Royal College of Physicians of Edinburgh in 1694.

9. D. Gregory, who left the University of Edinburgh for Oxford in 1691, having been harassed by a vehemently presbyterian university visitation.

10. The following account of the cases of the Scottish Enlightenment summarises Phillipson, "Culture and Society", and draws on Christie "Scottish scientific community".

11. This oscillation was particularly noticeable in the activities of A. Ramsay's Easy Club. Christie, ref. 1, p. 125.

12. For details of the University of Edinburgh's reform with regard to science, see Christie, ref. 1, 124–31.

13. C. Maclaurin (1748), *An Account of Sir Isaac Newton's Philosophical Discoveries* (London), 3–5, 390–2.

14. See Davie (1965), "Berkeley's impact on Scottish philosophers", *Philosophy*, **xl**, 222–34, for the importance of the Rankenians to the development of Scottish philosophy.

15. Christie, ref. 1, 127–8.

16. The Edinburgh lawyers could be characterised as the professional wing of landed society. Between 1707–51, 96% of Faculty of Advocate entrants came from a landed background, and between 1752–1811, 88%. N. T. Phillipson (1967), "The Scottish Whigs and the reform of the Court of Session, 1785–1830" (University of Cambridge Ph.D. dissertation).

17. Hume (1932), to A. Ramsay jnr., April 1755. *The letters of David Hume* (ed. J. Y. T. Greig), 2 vols. Oxford, **i**, 219–20.

18. For a forthright expression of economic responsibility by a member of the gentry on his class's behalf, see Sir John Clerk of Penicuik (1730), "Observations on the present circumstances of Scotland". Scottish Record Office, Clerk of Penicuik muniments, MS 3141 (ed. T. C. Smout) and published in *Miscellany of the Scottish History Society*, **x** (1965), 177–212.

19. Phillipson, "Public opinion and the union in the age of association", Phillipson and Mitchison, ref. 5, 142–3.

20. The term "neo-Harringtonian" was coined by J. Pocock. For further discussion of Harringtonian elements in the eighteenth century, see Pocock (1965),

"Macchiavelli, Harrington and English political ideologies in the eighteenth century", *William and Mary Quarterly*, 3rd series, **xxii**, 549–83.

21. A. Fletcher (1732), *Political Works of Andrew Fletcher*, London. Clerk, op. cit. Hume (1742), *Scots Magazine*, **iv**, 119.

22. For further details of the improvers' activities, see Smout (1969), *A History of the Scottish people, 1560–1830* (London), 291–301, and "The Landowner and the Planned Village", Phillipson and Mitchison, ref. 5, 73–106.

23. Davie, ref. 14.

24. This is a point well taken by H. T. Buckle (1970), *On Scotland and the Scotch Intellect* (ed. H. Hanham), (Chicago), 294–5.

25. H. Home (1754), "Of the laws of motion", *Essays and Observations, physical and Literary, read before a society in Edinburgh and published by them*, **i**, 1–69. John Stewart, "Some remarks on the laws of motion, and the inertia of matter", ibid, pp. 70–140. Hume's tactful editing defused a potentially divisive confrontation. See his letter of February 1754 to J. Stewart, in Greig, ref. 17, 185–88.

26. Cantor, "The Academy of Physics".

27. That this was a pretext is persuasively documented by Morrell (1975), "The Leslie Affair", *Scottish Historical Review*, **liv**.

28. G. Berkeley (1734), *The analyst* (London). C. Maclaurin (1742), *A treatise of fluxions* (Edinburgh).

29. E.g. T. Reid (1818), *Inquiry into the human mind* (Edinburgh), 172–203.

30. J. D'Alembert (1963), *Preliminary discourse to the Encyclopaedia of Diderot* (transl. Richard Schwab, New York), 16–21.

31. *Lectures on the elements of chemistry, by Joseph Black, M.D.* (ed. J. Robison), 2 vols., Edinburgh, 1803, **i**, 547–8. Robison also claims here that Black's doubts about Lavoisier's system of chemistry stemmed from precisely analagous philosophical and pedagogic preferences.

32. *Scots Magazine*, **v** (1743), 385.

33. L. Dobbin (1936), "A Cullen chemical manuscript of 1753", *Annals of science*, **i**, 140–2. William Cullen, Royal College of Physicians of Edinburgh MS C12 (student lecture notes, 1757–8).

34. Dobbin, ref. 33, p. 142–3.

35. J. Thomson (1859), *An account of the life, lectures and writings of William Cullen, M.D.* (2 vols., Edinburgh and London), **i**, 118–9.

36. See Hume's letter to Cullen, 21 January, 1752. Greig, ref. 17, p. 163.

37. For the Cullen–Kames correspondence on scientific agriculture, see Thomson, ref. 35, pp. 62–4, 592–601.

38. A. and N. Clow (1952), *The chemical revolution* (London), 177.

39. For the importance of Argyll to the success of Cullen's candidature, see Thomson, ref. 35, pp. 88–90. Details of their interest in salt purification are contained in the Cullen–Kames correspondence, ref. 35, pp. 598–600.

40. The Scottish investigation of pneumatic chemistry was continued by Black's pupil Daniel Rutherford. Black's work on heat was extended by J. Watt and W. Irvine.

41. J. Black, autobiographical note, Edinburgh University Library MS Gen. 874.

42. Clow and Clow, ref. 38, p. 177.

43. Kames to Cullen, 5 January 1750, Thomson, ref. 35, p. 593. Cullen to Kames, 17 January 1750, ref. 35, p. 595.

44. J. Black (1771), "An explanation of the effect of lime upon alkaline salts; and a method pointed out whereby it may be used with safety and advantage in bleaching", in Francis Home, *Experiments on bleaching* (2nd ed., Dublin), 265–82.

45. L. Laudan (1970), "Thomas Reid and the Newtonian turn of British method-

ological thought", in *The methodological heritage of Newton* (eds. Butts and Davis, Oxford), 103–31. Cantor, "Scottish methodological tradition".

46. *Medical essays and observations, revised and published by a society in Edinburgh* (5 vols., Edinburgh, 1733–44), **i**, p. xvi.

47. Christie, ref. 1, p. 132.

48. A. Smith (1795), "The principles which lead and direct philosophical enquiries: illustrated by the history of astronomy". This essay was probably written in the early 1750s, but did not appear until after Smith's death, when together with other early works it was edited by J. Black and J. Hutton and published in *Essays on philosophical subjects, by the late Adam Smith, LL.D.* (London). For Smith's philosophy of science, see pp. 3–26.

49. Reid, *Inquiry*, quoted by Buckle, ref. 24, p. 297.

50. Reid, *Inquiry*, ref. 29, pp. 129–30.

51. Reid, *Inquiry*, ref. 29, pp. 25–6.

52. Cantor, "The reception of the way theory of light in Britain: a case study illustrating the role of methodology in scientific debate", *Historical studies in the Physical sciences* (forthcoming).

53. Phillipson, "The Scottish Whigs".

54. Shapin, "The founding of the Royal Society of Edinburgh". Cantor, "Academy of Physics".

55. Smith (1896), *Lectures on justice, police, revenue and arms* (ed. Edwin Cannan, Oxford), 255–7.

56. J. D. Forbes (1862–6), "Opening address, session 1862–3", *Proceedings of the Royal Society of Edinburgh*, **v**, 17–23. Brewster (1830), "Review of Charles Babbage's *Decline of science in England*", in *Quarterly review*, **xliii**, 324–5 (note).

57. For Scottish resistance to anglicising university reform, see Davie, *Democratic Intellect*, 26–102.

58. Robison, quoted by Olson, ref. 1, 42.

59. W. Hamilton, quoted by Davie, *Democratic Intellect*, 127.

60. Brewster, "Review of Babbage", 341.

61. From a traditional lament for Scottish heroes.

8. Scientific Careers in Eighteenth-century France*

R. HAHN

(*University of California, Berkeley, USA*)

It is generally assumed—and correctly in my opinion—that the scientific successes registered in Old Regime France were in large part conditioned by the existence of a set of government-supported institutions, notably the Academy of Sciences, the Collège Royal, the Paris Observatory, the School of Military Engineering at Mézières and the Jardin du Roi.[1] It is the object of this presentation to explore some of the mechanisms that induced the growth of science, paying particular attention to the transformation witnessed in the actual life of scientists—or to use a short-hand term, to their careers. When I set out to investigate this topic, I expected to elaborate further some notions set forth in the first five chapters of my book on the Academy of Sciences[2] on the disappearance of amateur science in the wake of the professionalism displayed in the Academy, and on the establishment of a well-defined community of scientists whose expression was found in the activities of that Academy.[3] As you will see, new and ongoing research has caused me to rethink the problem, to introduce some significant nuances, and perhaps raise some more general issues which the French case highlights. In the context of the entire symposium, we may want to formulate some new perspectives on the nature and timing of the social transformations in science which have led us to our present situation.

The scenario I had expected for this paper was relatively simple and runs as follows. Between the time of Louis XIV and the French Revolution, institutions were created or refurbished by the State, thus providing visible evidence of the public sanction of scientific activities, and offering the prospect of established scientific careers in well-paid positions. The existence of such opportunities, at the same time unequalled anywhere in the world, drew scores of gifted men into the scientific orbit, increased the available pool of inventive talent on which progress always depends, and created salutary pressures on the quality and size of the scientific community. Thus for both innovation and critical discussion, France profited greatly from government sponsorship of scientific institutions. To put it another way, the leading sector

* Another version of this paper is to be published in *Minerva*.

of the scientific economy was propelled by *institutionalisation* and *professionalisation*—two critical factors usually observed for the study of the economy of developing societies, but not generally considered in the realm of science. It seemed to be the case for French science that creativity and productivity were spurred on by these two social variables.

If the assumption was valid, there would be need to concentrate on the mechanism by which these factors acted and to understand the motives that lay behind both the strengthening of institutions of science and the transformation of an informal community of scientists into a coherent social grouping. My research task would involve uncovering the motives behind the increasing State support of science, examining the government's response for assistance from individual scientists and ferreting out the reasons for the government's preference for organised and centralised institutions to supervise scientific activities.[4] It would also require a description of the gradual emergence of a new social class or occupational group, to measure its size over time, to examine its insertion into the national scene and to analyse expressions of its ideology. My expectations were that the eighteenth century was a critical period for this kind of social transformation in France.

Having worked a good deal on institutions, I decided to concentrate at first on scientists, and to look at manifestations of them as a class. At the outset, some very troublesome problems occurred, which I now judge to be symptomatic of a deeper issue. The word "scientist" in French, *un scientifique*, used as a substantive, does not occur commonly until the twentieth century. The translation into French of the term "natural philosopher", used so much in England during the eighteenth century, yields a notion closer to the German *Naturphilosoph* than desired, leading us away from the idea of a professional specialist. The most congenial contemporary term is *savant*, which often overlaps with *érudit*, implying a person of great knowledge.[5] In the eighteenth century, and in common parlance today, this appellation encompassed historians, antiquarians, numismatists, archaeologists and others, but not all persons we would wish to be included as scientists. For example, the creator of automatons Vaucanson, an important scientist by most definitions, would never have been called a *savant*.[6] It seems as if that term does not fit exactly with our notion of *un scientifique*.

But even if *savant* is adopted in order to avoid glaring anachronisms a major problem still remains. The category of *savant* refers to a mental activity, at times to a type of social behaviour, but scarcely to a social or occupational class. When looking at a series of notarial archives, as I have done recently, at marriage contracts, property sales, successions; or at tax records and payrolls, where one finds occupational labels of all sorts, the word *savant* itself never appeared. Most often the individual we know to have been a scientist is listed as a member of a learned society, as a professor, as a doctor, as an army officer, as a clergyman. Occasionally the term *homme* or *gens de lettres* is used, especially for those who published textbooks, dictionaries or edited journals.[7]

This linguistic difficulty is disturbing not only for the would-be quantitative social historian who needs well-defined categories to make some respectable and valid headcounts. It also bothers the general historian who might reasonably expect a close association between the emergence of science as a professional activity in the late seventeenth century and the establishment of a recognisable social group devoted to its pursuit. The absence of a single, unequivocal term to describe this group is an indicator of a great complexity in the social situation. The likelihood that such terminological confusion reigned in France well into *this* century (in a country that prides itself upon attention to its language) is even more alarming for the historian seeking generalisations about the social evolution of the life of science.

How then should one proceed? I took as my field of study the more than 300 working members of the Academy until the Revolution, and attempted to determine to what extent they could be said to be part of the class of scientists. Ideally, one would want to know what values they shared, what their self-image was, how they made a living, what degree of social cohesiveness they displayed, and what career patterns their biographies reveal. The full answer to these questions is still pending, but I want to offer some partial results.

While they formed a scientific community in the intellectual sense—that is, sharing values about the use of reason, observation, experimentation and objectivity to uncover progressively natural reality—while they communicated with each other, sought and accepted peer review and agreed to act as public arbiters of science in France, they did not constitute a homogeneous class.[8] If we knew with more assurance their social origins, we would have one element explaining their lack of cohesion. But more importantly for our purposes, nothing in their professional lives as academicians forced them into a single mould. Entrance to the community was based on merit rather than social origin, though in practice lower orders were surely excluded. Illiteracy was a bar, but as long as it could be surmounted, it mattered little before the Revolution if the academician was self-taught, apprenticed or formally schooled. There was no religious test as such peculiar to academicians, the only restriction being that members of regular orders were ineligible for full pensions.[9] This did not exclude several clergymen who acted as full-time scientists: the abbés Nollet, La Caille and Haüy, to name only the most famous. But their bond to the religious life was weaker than that of Jesuits or Oratorians. Nor were members of guilds excluded because of their affiliations. Plenty of academicians remained pharmacists or surgeons. The one notable exception to the rule was the clock-maker Jean-Baptiste Le Roy who explicitly left his occupation upon election.[10] In his case, I suspect the operating principle behind the desired withdrawal from clock-making was the fear that he would be suspect when acting as a judge of technical inventions brought before the Academy. The intent of the ban was similar to the semi-exclusion of members of the regular orders: to avoid double allegiances which would conflict with a scrupulous adherence to impartiality. In any case, there was no intrinsic social contradiction between practising a craft, operating a business or holding a benefice and being an academician. We even have on record the bankruptcy of one academician, Quatremère d'Isjonval.[11]

The collective life of the academicians was not a source for much cohesiveness either. Outside the bi-weekly meetings lasting for two hours each, occasional committee gatherings, a yearly mass celebrated at the Church of the Oratoire, and two public assemblies, there were no other collegial activities to bind them together. So far as is known, they did not act as a large group socially. On the contrary, the annals of the Academy are replete with bitter cleavages. Factionalism was rampant, and manifested itself ferociously on the occasion of elections, and this divisiveness was further aggravated by political differences as the Revolution broke out. There were cliques and schools, often based on specialities within science or on conflicting approaches to scientific questions, and reinforced by patronage links. I was also struck in my search through notarial records by the relative rarity of marriages between academicians' relatives, and how seldom academicians were witnesses to each others' weddings. A few well-known dynasties of scientists existed (the Cassinis, the Geoffroys, the Delisles, the Jussieus and the Le Monniers) but family connections had less to do with the life of the Academy than with the practice of handing down certain non-academic offices from one generation to the next, as was common in the Old Regime. To finish the list of centrifugal forces in the academic community, one should add the wide disparity of social class within the membership. Some issued from modest origins—Antoine Petit or Gaspard Monge—others were in the administrative or parliamentary milieus—Fantet de Lagny, Grandjean de Fouchy, Mignot de Montigny, Dionis du Séjour—while certain academicians were or became nobles—Cassini, Quesnay, Courtivron, Chabert, Perronet.[12]

It is reasonable to expect that improving financial opportunities in science would counteract these centrifugal tendencies by creating an occupational class of full-time scientists. In view of the professionalism displayed in the Academy, it would be natural to expect that academicians would be the first to fit into or even to create this occupational group. They were singularly devoted to high standards in science that set them off from laymen or amateurs; they were generally self-motivated and self-regulated; and they performed specialised tasks in French culture that were recognised and sanctioned by the State.[13] Another step toward the full professionalisation of scientists could easily have been taken if salaries had been assigned to academicians, thereby offering a clearly-defined and desirable career through membership in the Academy.[14] But here too, there seem to have been some complications that prevented the establishment of a fully recognisable social group of scientists. The financial arrangements in the Academy were such that on the eve of the Revolution, a young scientist could not expect to make a living directly from belonging to the learned society. To understand this peculiar situation requires a close look at academic financing.

In the seventeenth century, Louis XIV's ministers paid individual academicians directly. Most received an annual sum of around 1500–2000 French livres from the Crown, while some foreigners like Huygens and Cassini were kept with considerably higher sums (6000–9000 livres) necessary to attract them to France and keep them there. The budget for all these scientists (which I have anachronistically extracted from payrolls for all kinds of employees)[15] hovered around 30 000 livres in the 1670s. A century later, academic salaries, now handled through the institution, totalled around 50 000 livres annually, 6000 for each of the six academic classes, 3000 each for the Secretary and Treasurer, and 8000 for the miscellaneous *petites pensions*.[16] In the meantime, the membership eligible for pay had jumped from approximately 20 to 50, and some inflation had further reduced the value of average academic salaries. On the average, academicians fared better financially under the Sun King than under his two royal successors despite the increasing budget for science.

From our point of view of scientific careers, there is an even more serious issue residing in the distribution of funds. They were in fact more like old-age pensions than salaries. For most of the eighteenth century, the Academy was divided into six classes according to scientific discipline, and within each class there were three ranks: two *élèves* or *adjoints*, three *associés* and three *pensionnaires*. Because promotion was a matter of seniority, and generally within the same class, a young academician had to wait until five of his elders passed on before having a share of the regular pensions for his specialty. And since these pensions were graduated, with the elder receiving 3000 livres annually, while his two colleagues collected 1800 and 1200 livres, only at a ripe old age could a scientist expect to earn a living directly and solely from being a member of the Academy.[17] The mathematician Etienne Bézout waited twenty-four years before receiving his meagre 1200 livres stipend, and died the year after of old age. Legentil waited thirty-two years, until he was sixty before being on the academic payroll.

A serious gap existed between what historians refer to proudly as funded government sponsorship of French science, and the life of the individual scientist. Talented young minds could not objectively have been attracted into science for pecuniary reasons, as was the case in the United States after World War II. Indeed, if they read eulogies written by the Academy's Secretaries, they would have been forewarned. Fontenelle, in speaking of Rolle's career says bluntly that "there is between science and wealth an old and irreconcilable distinction".[18] Condorcet, in explaining why Bézout's family was reluctant to allow him to pursue his scholarly vocation indicates that "a father . . . knows that education and enlightenment lead neither to honour nor to fortune".[19] Another Secretary, Grandjean de Fouchy, insists more explicitly that the function of eulogies is in part to display for the public by example all the sacrifices required for those having dared to make a career out of science:

> "The history of the Academy . . . teaches how one can overcome difficulties in the study of science itself, but those who have the courage to enter this laborious career (*laborieuse carrière*) meet up with obstacles of another kind as well, they must be ready to surmount hurdles placed in their way, even by those closest to them. The Academy requires qualities of will as well as of mind, and both must be worked at. There is no better way to instruct the young wanting to cultivate science than by recounting the lives of late academicians, giving a summary of their life, work, and the way which they turned their difficulties. Not only is this a way to pay them the tribute they deserve, but also to offer for those who may one day pursue this type of occupation (*genre d'occupation*) an example of the honours they may expect at the end of their career, and a model of ways to overcome the obstacles they would likely encounter."[20]

The picture of the self-sacrificing scientist as a struggling hero is not merely a rhetorical device used to good effect. Fouchy, like Fontenelle and Condorcet, is also mirroring the tensions involved and contradictions between the emotional values derived from a scientific career and the low material rewards that flow from it. In France, the research activity, satisfying as it might be for the soul, was by no means a path to wealth or even social esteem. To a large extent that statement still applies in our own times.

All this surely helps to explain why academicians did not constitute an occupational class, but it leaves unanswered the question of how those few courageous enough to engage in scientific activities made ends meet, and what the implications of such a situation might be.

You may have noticed that my reference to stipends was to earnings directly dependent upon membership in the Academy. The prestige of belonging to the learned society and the contacts made by academicians in the line of duty were of capital importance for earning what was in fact the major source of income for them. We might be tempted, therefore, to include these earnings in our calculation of the benefits deriving from membership. The connexion undoubtedly existed, but to lump these salaries together would mask the issue to be discussed later, whether these academic byproducts reinforced or detracted from the furtherance of careers in science. Before turning to that issue, let us categorise the types of positions outside the Academy where me might find our academician earning a decent living.

Teaching constituted the most frequent and congenial of jobs held by academicians. They dominated in two institutions older than the Academy, the Collège Royal and the Jardin du Roi. In the former, one finds mostly mathematicians, astronomers and medical men.[21] At the Jardin du Roi, naturalists able to sell their science to students of medicine, pharmacy and surgery also tended to be academicians, or used the posts as stepping stones to join its ranks.[22] There were also accredited professors at the Faculty of Medicine and the Colleges of Pharmacy and Surgery.[23] Mathematicians found military schools, the civil engineering corps and the architectural and surveying professions interested in their specialty. When student examiners' posts were created, they were often attributed to qualified mathematicians.[24] In addition, Paris abounded with ephemeral paying public courses or demonstration lectures that brought money from its auditors. Success for these enterprises, however, depended more on eloquence and notoriety than professional competence, which was the trademark of academicians. Some like the abbé Nollet, Rouelle and Fourcroy were nonetheless very successful financially in this popular role. Quieter scientists preferred employment as tutors in rich families or at Court.

A second category of employment was as directors of royal scientific establishments, or as scientists to the Court. Chirac, Dufay and Buffon earned a good living from being Intendant of the Jardin du Roi; the Cassini dynasty ran the Observatory, and found some money to employ some of their colleagues as researchers. La Peyronie and Quesnay directed the Academy of Surgery, while Lassone and Vicq·d'Azyr lorded over the Society of Medicine, all with good stipends. The King, Queen and members of the Royal family each had several physicians, apothecaries and surgeons, many from the academic establishment.[25] There was a post as Royal Geographer, a keeper of natural history collections for the Duc d'Orléans, and an impressive list of secretaries, tutors and health specialists in the households of King Louis XVI's brothers as well.

We find academicians in a third and related sector, on military payrolls. In addition to the professors and examiners of officers' candidate schools, the Navy had a section for maps and charts, and another for naval engineering run by academicians. The *Connaissance des Temps* was entrusted to astronomers from the Academy. Duhamel du Monceau, Le Monnier, Lalande, Deparcieux and Rochon, among others, figured on naval payrolls as well.[26] On the Army side, improving fortifications and making maps were activities also yielding stipends for scientists.[27] Moreover, the Academy had several prominent officers in its ranks who drew regular salaries: Bélidor and the chevalier d'Arcy; Deschiens de Ressons and Vallière in the artillery; Fourcroy de Ramecourt, Coulomb and Meusnier in the engineering corps; and La Galissonnière, Borda and Bory in the navy. The list is surely not complete.

The fourth and most novel category, both for its character and significance for academicians who were neither mathematicians nor in the healing arts was as government consultants in technically-oriented enterprises.[28] Tillet, d'Arcet and Condorcet, following Newton's footsteps, were at the Mint; Réaumur, Hellot, Morand, Duhamel and Sage found places in mining and metallurgy; others were consultants for the Gobelins tapestry works or at the Sèvres porcelain manufacture, both government enterprises. Lalande, d'Alembert and Bossut advised the State on inland navigation.[29] And there were numerous inspectors of commerce and industry, including Dufay, Hellot, Montigny, Macquer, Vaucanson, Desmarest, Jars, LeRoy, Vandermonde and Berthollet.[30] Some even engaged in industrial spying for the administration, for which they were paid probably from secret coffers. Finally, the government used academic manpower to censor the book trade in scientific and technical domains.

One must not forget academicians who were engaged in activities quite distant from their scientific specialties, like Lavoisier who was a tax collector.[31] Mignot de Montigny was Treasurer of France; Fantet de Lagny was assistant director of a government bank; d'Onsenbray was postmaster-general; and Dionis du Séjour a prominent legal counsellor. A few academicians who inherited their wealth tended to their land and investments, though they were rarer than one might expect. Of the most significant among the active scientists were the Duc de Chaulnes and the Marquis de Courtivron.

The account given above, incomplete as it is, gives the impression of a vast infrastructure of positions in Old Regime society which could support in a reasonable way those engaged in scientific pursuits, and permit them to live often with some comfort, leaving a decent inheritance for their children. While it helped individuals and their families, can we also say it supported the pursuit of science as a research activity? The answer is complex and difficult, because it requires us to think carefully about what we mean by science.

On the positive side, especially when compared to the situation of scientists in other lands, the French academician was quite fortunate. The miscellaneous posts listed above brought in varying amounts of money, ranging from 1800 livres annually as a professor at the Ecole royale Militaire in Paris to 6000 livres as examiner of naval schools. For some budding scientists, like Laplace for example, it meant the difference between taking holy orders, thus entering a tried and tested career, and following his inclination to be a scientist. Once embarked in this novel line of work, it was rare to see a promising and talented young man revert to a more conventional occupation.

But there were drawbacks as well. Since jobs were inserted marginally into the Old Regime occupational structure, and were at the mercy of individuals whose commitments were not to science, and since the availability of funds and posts was known only privately, the whole network was extremely fragile. Policy changes based on political preferences could upset the entire system. Moreover, scientists were put in a position of periodically pleading their case to higher authorities who might be ignorant of the content of science and its practitioners' aspirations. Fortunately for men living in the Age of

Enlightenment, the public image of science was positive. The same could not be said for the following century.

There was an even greater problem. Most of the positions cited above involved teaching the elements of science, or applying knowledge of nature to practical problems. They did not generally encourage scientists' desire for the advancement of science, and rarely provided the time or means to extend the limits of the research frontier. In a few cases, the occupation stimulated some significant new research in the Sèvres manufacture, in the preparation of nautical almanacs, or on experiments with the shape of ship hulls. Most often, academicians were employed to use the by-product of their scientific excellence, to be useful to society. But between research and utility, there could be, and were, conflicts of time, of interest, even of method. D'Alembert recommended a teaching post to Laplace saying that it required giving classes only in the mornings, leaving the rest of the time, it was understood, for real science![32] Mignot de Montigny, who spent his whole career applying science to practical issues, writing but one published research paper, felt sufficiently torn about the matter to leave a twelve-page explanation of his conduct as an appendix to his last will.[33] Macquer, in a letter to his wife, recounts how uncomfortable he was waiting for his administrative masters to emerge from their councils, yet recognised it was necessary to waste his time paying court to them.[34] Other examples could surely be cited.

One thing seems clear, and requires emphasis. The spirit of research for the furtherance of rational understanding about nature (which is my definition of scientific activity) neither coincided completely with the needs of Old Regime society, nor was encouraged on the scale required to create a socio-professional class of scientists. Even in the Academy, which fostered the purest kind of research, there was a demand on the members to act as government consultants which detracted from the ideal set forth above. The financial necessities of scientists required them to be pulled in different directions, and to be turned away from the calling of science. They acted out roles as teachers of laymen, consultants and administrators without reinforcing their roles as researchers. Thus, instead of the crystallisation of the scientist as an acceptable and autonomous occupation, our eighteenth-century academician experienced the centrifugal forces that tempted him in other directions. For some, science may only have been a way-station on the road to fame or power.[35] If despite the advances made in the French Revolution in the institutionalisation of certain scientific roles, nineteenth-century scientists experienced serious difficulties that shook their will to advance the limits of our understanding, it was to a large extent because the career of research scientists failed to emerge clearly during the Age of Academies. The modern scientist was still in the making.

Some comments of a more general nature may be in order before leaving the subject. Though France led the way in the Enlightenment by making the practice of science an acceptable activity and by honouring its practitioners, it did not provide the model for the modern professional scientist. There was a noticeable lag between the professional manner in which science was practised and the creating of an occupational class devoted principally to its progress. While the philosophic ideology of the Scientific Revolution had shaped the self-sustaining and self-correcting activity we perceive as an early form of modern science, the development of an associated occupational category operating in as autonomous a fashion as other professions was barely beginning to emerge. Even before the Revolution created a standardised system of higher education in France, careers in science were by necessity anchored in cognate activities that rarely focused on a set of common problems on the frontier between the known and the unknown.[36] Careers were not made by concentrating on clustered research topics. It was this "dispersal" more than the scientistic outlook[37] that became a habit among those engaged in scientific research, and ultimately made it impossible for the French to meet the German and English challenge in the course of the nineteenth century. Because of the career patterns forged in the eighteenth century, the advancement of science remained a part-time activity well into our own century even for those who were trained by the most progressive educational system in the world.

References

1. For bibliography on French science before the Revolution, see *Enseignement et Diffusion des Sciences en France au XVIIIè Siècle* (ed. R. Taton, Paris, Hermann, 1964); S. L. Chapin (1968), "The Academy of Sciences during the Eighteenth Century: An Astronomical Appraisal", *French Historical Studies*, **v**, 371–404; M. P. Crosland (1973), "The History of French Science: Recent Publications and Perspectives", *French Historical Studies*, **viii**, 157–171; and R. Taton (1973), "Sur Quelques Ouvrages Récents Concernant l'Histoire de la Science Française", *Revue d'Histoire des Sciences*, **xxvi**, 69–90.

2. R. Hahn (1971), *The Anatomy of a Scientific Institution: the Paris Academy of Sciences, 1666–1803* (Berkeley, University of California Press); and the critical reviews of M. P. Crosland (1972), *Isis*, **lxiii**, 405–7; K. M. Baker (1972), *Minerva*, **x**, 502–8; H. Brown (1972), *Annals of Science*, **xxix**, 313–6; and J. Ben-David (1972), *Journal of Modern History*, **xliv**, 589–92.

3. R. Hahn (1968), "The Problems of the French Scientific Community, 1793–1795", *Actes du XIIè Congrès International d'Histoire des Sciences* (Paris), **iiib**, 37–40.

4. R. Rappaport (1969), "Government Patronage of Science in Eighteenth-century France", *History of Science*, **viii**, 119–36; and S. L. Chapin (1968), "Scientific Profit from the Profit Motive: The Case of the La Pérouse Expedition", *Actes du XIIè Congrès International d'Histoire des Sciences* (Paris), **xi**, 45–9.

5. For the period up to 1700, see U. Ricken (1961), *"Gelehrter" und "Wissenschaft" im Französischen. Beiträge zu ihrer Bezeichnungsgeschichte vom 12.–17. Jahrhundert* (Berlin, Akademie Verlag); and S. Ross (1962), "Scientist: The Story of a Word", *Annals of Science*, **xviii**, 65–85.

6. A. Doyon et L. Liaigre (1966), *Jacques Vaucanson, Mécanicien de Génie* (Paris, Presses Universitaires de France).

7. P. Bénichou (1973), in *Le Sacre de l'Ecrivain 1750–1830* (Paris, Corti), 23–77 discusses the analogous problem of the self-image of the French writer during the Old Regime. See also M. Gaulin (1972), "Le Concept d'Homme de Lettres, en France, à l'Epoque de l'Encyclopédie", Ph.D. dissertation, Harvard University.

8. Despite their allegiance to the Academy, they did not form a recognisable class of academicians either. Members of the three major royal academies in Paris (Académie Française, Académie des Inscriptions et Belles-Lettres and Académie des Sciences) cohered even less than scientists. Only a few (Fontenelle, Buffon, d'Alembert, Bailly, Condorcet) looked upon their title of academician as a career goal, and they were exceptional individuals who displayed a strong scientistic bent. See also D. Roche (1974), "Sciences et Pouvoirs dans la France du XVIIIè Siècle (1666–1803)", *Annales, Economies, Sociétés, Civilisations*, **xxix**, 746–8.

9. See article XII of the 1699 regulations of the Academy, modified by the 1716 letters-patent, in *L'Institut de France. Lois, Statuts et Règlements Concernant les Anciennes Académies et l'Institut de 1635 à 1889* (ed. L. Aucoc, Paris, Imprimerie Nationale, 1889), lxxxvi and xciii.

10. Undated letter of Le Roy to Mairan (shortly before 18 August 1751), in Archives de l'Académie des Sciences, dossier Le Roy, Jean-Baptiste, reprinted in part by J. L. F. Bertrand (1869), *L'Académie des Sciences et les Académiciens de 1666 à 1793* (Paris, Hetzel), 74–5.

11. Archives de Paris, D 4B[6] 107 (7585).

12. There is no study of the social composition of academicians available. Series DC[6] at the Archives de Paris indicates that in addition to the nobles listed, Bouvart, Portal and Tillet were ennobled while already academicians.

13. These features are generally included in the characterisation of "professionals", particularly scientific professions. See J. Ben-David (1972), "The Profession of Science and its Powers", *Minerva*, **x**, 362–83. See also N. Reingold, "Definitions and Speculations: The Professionalisation of Science in America in the Nineteenth Century", read before the American Academy of Arts and Science's Conference on Early History of Societies for the Promotion of Knowledge in the United States held in summer of 1973 (in press).

14. R. M. MacLeod (1972), "Resources of Science in Victorian England: The Endowment of Science Movement, 1868–1900", *Science and Society 1600–1900* (ed. Peter Mathias, Cambridge, Cambridge University Press), 111–66.

15. *Comptes des Bâtiments du Roi sous le Règne de Louis XIV* (ed. J. J. Guiffrey, Paris, Imprimerie Nationale, 1881–1901), 5 vols.

16. "Journal du Trésor de l'Académie, 1792–1793", fol. 3, Manuscript 114, Cornell University.

17. I estimate that one could live modestly in mid-eighteenth century Paris with an annual revenue of 5000 livres.

18. B. de Fontenelle (1825), *Oeuvres* (Paris, Salmon), **ii**, 25.

19. M.-J.-A.-N. Caritat de Condorcet (1847–9), *Oeuvres* (Paris, Didot), **iii**, 43.

20. J.-P. Grandjean de Fouchy (1761), *Eloges des Académiciens de l'Académie Royale des Sciences, Morts depuis l'An 1744* (Paris, Brunet), **i**, iv–v.

21. L. A. Sédillot (1869), *Les Professeurs de Mathématiques et de Physique Générale au Collège de France* (Rome, Imprimerie des Sciences Mathématiques et Physiques); A. Lefranc (1893), *Histoire du Collège de France depuis ses Origines jusqu'à la Fin du Premier Empire* (Paris, Hachette).

22. J.-P. Contant (1952), *L'Enseignement de la Chimie au Jardin Royal des Plantes de Paris*

(Cahors, Coueslant); P. Crestois (1953), *L'Enseignement de la Botanique au Jardin Royal des Plantes de Paris* (Cahors, Coueslant).

23. P. Delaunay (1906), *Le Monde Médical Parisien au Dix-huitième Siècle* (Paris, Rousset); P. Huard (1967), *L'Académie Royale de Chirurgie* (Paris, Palais de la Découverte); M. Cazé (1943), *Le Collège de Pharmacie de Paris (1777–1796)* (Fontenay-aux-Roses, Bellenand).

24. D. I. Duveen and R. Hahn (1957), "Laplace's Succession to Bézout's Post of Examinateur des Elèves de l'Artillerie", *Isis,* **xlviii**, 416–27.

25. Rules of the Academy prevented scientists living in Versailles from holding regular academic posts without special dispensation. See Aucoc, ref. 9, pp. lxxxiv–v and examples of dispensations in Manuscript Ashburnham 1700, Biblioteca Medicea-Laurenziana, Florence, p. 32–3.

26. For details, see C²50–51 of the Fonds Anciens de la Marine, Archives Nationales.

27. H. M. A. Berthaut (1902), *Les Ingénieurs Géographes Militaires 1624–1831* (Paris, Imprimerie du Service Géographique de Armée), 2 vols.

28. H. Guerlac (1959), "Some French Antecedents of the Chemical Revolution", *Chymia,* **v**, 72–112.

29. *Correspondance Inédite de Condorcet et de Turgot 1770–1779* (ed. C. Henry, Paris, Charavay, 1882), 260 *et seq.* discusses the establishment of a non-salaried committee on inland navigation. The scientists were rewarded in other ways, as discussed in R. Hahn (1962), "The Chair of Hydrodynamics in Paris, 1775–1791: a Creation of Turgot", *Actes du Xè Congrès International d'Histoire des Sciences* (Ithaca), **ii**, 751–4.

30. F. Bacquié (1927), "Les Inspecteurs des Manufactures sous l'Ancien Régime, 1669–1791", *Mémoires et Documents pour Servir à l'Histoire du Commerce et de l'Industrie en France,* **xii**, 19–139.

31. The tensions caused by Lavoisier's occupation and his scientific interests are discussed by M. Vergnaud (1954) in "Un Savant pendant la Révolution", *Cahiers Internationaux de Sociologie,* **xvii**, 123–39. A recent article by L. Scheler (1973), "Lavoisier et la Régie des Poudres", *Revue d'Histoire des Sciences,* **xxvi**, 194–222 nonetheless underscores that his well-paid government post in the powder manufacture was closely connected to his chemical activities.

32. Letter of d'Alembert to Le Canu dated 25 August 1769, Collection Historique, Institut Pédagogique, Paris.

33. 6 May 1782, Etude XCIX, liasse 669, Minutier Central des Archives Notariales, Paris.

34. Letter dated 19 October 1775, Manuscript fr. 9134, fol. 102, Bibliothèque Nationale, Cabinet des Manuscrits.

35. The career of Lacepède is a case in point. See R. Hahn (1974) "Sur les Débuts de la Carrière Scientifique de Lacepède", *Revue d'Histoire des Sciences,* **xxvii**, 347–53; and "L'Autobiographie de Lacepède Retrouvée", *Dix-Huitième Siècle* (in press).

36. The cases of Lavoisier's group at the Arsenal and Laplace's disciples in the Société d'Arcueil are notorious counter-examples which I judge to be important, but short-lived exceptions. They did not in any case thrive because of sustained government patronage, depending instead upon the personalities of their leaders and their financial backing. On this point, I follow the interpretation of R. Fox (1973), in "Scientific Enterprise and the Patronage of Research in France 1800–1870", *Minerva,* **xi**, 442–73 rather than M. P. Crosland (1967), in *The Society of Arcueil. A View of French Science at the Time of Napoleon I* (London, Heinemann).

37. J. Ben-David (1970), "The Rise and Decline of France as a Scientific Centre", *Minerva,* **viii**, 160–79.

9. The Development of a Professional Career in Science in France*

M. P. CROSLAND
(*University of Kent*)

A professional career entails full-time and remunerated employment entered into after a course of training.[1] Such careers and the prerequisite training became available in the educational structure of post-revolutionary France. They constituted a crucial phase in the establishment of science as a profession. The large number of scientific posts in official institutions in the early nineteenth century and the eminence of the men who filled them gave France unparalleled distinction in most branches of the physical and biological sciences. France was envied by men of scientific bent in other countries where there were neither educational facilities nor a significant number of posts which could employ scientific talents. The education and the mode of recruitment of scientists cannot be understood by confining attention to the study of any one institution or type of institution such as the university. Universities in eighteenth-century France were moribund and the national university established in 1808 with various faculties of science came only gradually to embrace the new professional scientists. Universities did not play a great part in laying the foundations for the profession of science in France.

One of the secondary characteristics of a profession is a claim to high standards of competence, standards imposed by the profession itself. If one remembers that the Paris Académie des Sciences had begun to impose certain standards under the ancien régime,[2] one naturally thinks of its successor, the First Class of the Institut, as the body which would continue and extend this tradition into the nineteenth century. Undoubtedly the Institut did include among its functions the imposition of standards, particularly for work for which it recommended publication, but because educational institutions became so important after the Revolution, it was only at the highest level that the Institut played a major role. From the point of view of education and entry to the profession, it was the Ecole polytechnique which played the most

* This article also appeared in *Minerva* (Spring 1975).

prominent role in the creation of standards, in the first place, standards of mathematical competence required for entry[3] but more pertinently for the standard of attainment in physical science and mathematics required for graduation. Both the staff and the governing body (the "conseil de perfectionnement") were influential here. A major influence was that of the final examiners.[4] In a system which started with highly selected students, who were subjected to an intensive course of instruction and frequent tests, it was the external examiners who put their vital seal of approval on the final stage of graduation. They exercised influence not only over the student body but over the teaching.

Once the Paris faculty of science got into its stride, it tended to take over from the Polytechnique, particularly in the 1820s and 30s. I would suggest that from the 1830s the doctorate became more and more the expected qualification for the aspiring academic scientist. The juries for the doctoral theses, drawn from the staff of the faculty and the *grandes écoles,* thus administered a diploma vital to the career of aspirants to the profession. The Ecole normale, too, played an increasingly important role in the training of scientists who would make a career in higher education.

The ancien régime

France in the eighteenth century was one of the most powerful states in Western Europe. Its population of about 28 million was nearly twice that of Britain. It inherited from Louis XIV ideas of a centralised state very far from the tradition of local autonomy common in Britain. The British traditions of independence and self-help[5] may be contrasted with measures of government control in France which affected both industry and science. The ideas of the French enlightenment brought to a focus the general French attitude that action should be based on reason, and practice should be based on theory. This respect for the intellect combined with the impossibility of a political career helped to turn many French minds towards science. A church which had lost power over intellectual activity, and the idealisation of the possibilities of science both contributed from different directions to produce a climate of opinion in which a new approach to nature was likely to flourish.

Much important science was done in France in the mid-eighteenth century, but a crucial change came with the Revolution when—if the event can be located at all precisely—science became a profession. This happened in several different ways; one path was by differentiation and specialisation. Under the *ancien régime* the man of science had hardly been distinguished from the man of letters. The terms *savant* or *philosophe* could be applied to both, and each, if fortunate, might receive a literary pension granted to *gens de lettres*. That there need be no sharp demarcation between science and literature is illustrated by the work of Buffon for whom style was as important as content.[6] Of course there were signs of a change even before the Revolution. As well as general books, men of science wrote technical monographs and memoirs more suitable for an academy than a *salon*. Of course the difference between science and letters was recognised by the existence of the Académie des Sciences as well as the literary Académie Française. Yet it was the Revolution and the war which forced a more absolute differentiation. Scientists provided quite different services for the revolutionary armies than did literary men. Playwrights could not advise on the casting of cannon; the imagination of poets was not of much use in inventions of war.[7] Science, or a certain interpretation of it, was also in the ascendant as a comprehensive outlook on life.[8] In the Napoleonic regime, science was specially favoured[9] at the expense of literature. This may be seen by a comparison of the treatment of Laplace, Berthollet and Cuvier with the greatest literary figures of the time: Chateaubriand, Benjamin Constant and Madame de Staël. One can detect signs of the beginning of a bifurcation of knowledge into "two cultures", officially recognised in the Napoleonic Université de France with its faculty of science on a parallel with a faculty of letters. Science now existed in the educational system in its own right and not as a part of an arts course or of philosophy or medicine. Although the *Journal des Savants*[10] continued to give news of science as well as literature, it was to the new specialised scientific journals, whether of mathematics,[11] physics and chemistry[12] or in the biological sciences,[13] that men would now turn for anything more than a superficial account of scientific developments. Specialisation was one of the features of science in early nineteenth-century France.

Even under the *ancien régime* work of a high standard was possible, as the achievements of Lavoisier clearly demonstrate.[14] But Lavoisier was not a professional scientist unless the word "professional" refers simply to work of a high standard. By other criteria, Lavoisier was an amateur, although this does not mean that his experiments were not most carefully planned, executed and reported. I prefer to use the term "professional" in an occupational rather than in an evaluative sense.

Salaries and Science

Lavoisier provides a convenient example of a stage in the growth of the profession of science. In the emergence of modern science, it was at first usual for men with scientific interests to be paid or employed to do something quite outside science. Lavoisier's occupation as a tax farmer related to science only insofar as he used his earnings to meet the expenses of his laboratory. It was a significant advance towards the full professionalisation of science when men could be employed to do work related to their scientific interests. The teaching of elementary science would provide a good example of this. In the *ancien régime* the mathematician Laplace improved his financial position by obtaining the post of examiner to the artillery. He explained that the great attraction of this post, apart from the salary was that the duties would not take up more than three to four weeks of the year.[15] Examining, like teaching, could make all the difference between a bare livelihood or none; it also was closely connected with science but it did not call for the performance of research as part of its stipulated duties.

The second stage in this simplified model of the establishment of science as a profession came when scientists were paid salaries to do scientific work which overlapped with the advancing frontiers of knowledge. They could be paid to do research, as at the Observatoire, or they might be given posts of advanced teaching.

The payment of an adequate salary was a significant step in the growth of the profession of science. In the first place it implies recognition of the value of the work done by a scientist. This is particularly important if the employer is not a wealthy private individual but a government. Condorcet was one of those who saw most clearly that with the Revolution the time had come for science to be a profession ("une sorte d'état") and not the leisure-time occupation of those for whom money was a matter of indifference.[16] With the establishment of the Institute in 1795, all members of the official body of science were for the first time paid salaries, and as a matter of principle they were bound to accept them.[17] There was no room in post-revolutionary France for the traditional wealthy amateur of the eighteenth century. Similarly, the abolition of the class of honorary members in the new Institute marked the end of a situation in which science could exist as an activity of a leisured nobility.[18]

The receipt of a salary implied a certain contractual obligation and a certain responsibility, whether to engage in the teaching of science or research or both. The most important feature of all about a scientist receiving an adequate salary *for his scientific work* was that it enabled him to pursue science as a full-time occupation which was recognised as such by the government.[19] This was a stage which was reached in the institutions established in France in the period of the Revolution.

Fourcroy, in one of his reports to the Convention, was quite explicit about the need to pay reasonable salaries to professors of science and medicine

"Your intention of reviving the useful sciences and of favouring their progress requires that professors and their assistants who have the responsibility of giving student's lectures both theoretical and practical should be attached exclusively to these duties and that no other private occupation should distract them from them. It is therefore necessary that their salaries should be sufficient for their needs and that they should not be obliged to look for other posts as a means of completing their livelihood. Men who have spent twenty years of their lives in study in order to acquire profound knowledge and be able to transmit it to others, should be treated by the country which employs them in such a way that they are not tormented by domestic anxieties. By the exercise of their useful talents they should be able to draw on resources sufficient for their own maintenance and that of their families."[20]

Teaching

One of the changes which took place in teaching at the time of the Revolution was the raising of the standard. Whereas an *abbé* under the *ancien régime* would have been glad to present the elements of science to boys in the final year of college and the mathematics teacher in the military academy would have helped cadets with their Bézout textbook,[21] a change came abruptly—perhaps too abruptly—with the Ecole normale of 1795 when the leading savants tried to present their own ideas on science.[22] It was impossible for the great mathematician Lagrange to come down to the level of his audience and the original ideas of Berthollet on chemical affinity and the nature of acids would not have been appreciated by the majority of his students. The Ecole, although short-lived in its first form,[23] provided an admirable precedent. It showed, among other things, that it was possible to establish educational institutions where the most learned scientists could present their own views of their respective subjects.

I distinguish here between teaching at an elementary level and teaching which overlaps with research, the former being today what one expects in a school and the latter what one hopes for in a university. The revolutionary situation in France enabled institutions to be established where teaching was at the level of research. Although Berthollet had little personal success as a teacher either at the Ecole normale or at the Ecole polytechnique, he was at least able to inspire his protégé Gay-Lussac[24] with a view of teaching as an extension of research. Even at the faculty of science, where teaching was not at a particularly high level, Gay-Lussac and Biot divided the physics course into two parts according to their research interests.[25]

These educational innovations had wider social implications both for teachers and students. We are all familiar with early nineteenth-century English novels where the governess, usually a gentlewoman by upbringing, is embarrassed by being treated as one of the servants. In eighteenth-century France there was some sort of parallel, although we should remember that the vast bulk of education was in the hands of the various religious orders. The social standing of the Oratorian or Jesuit was much less as a teacher than as a cleric.

The social position of a teacher depended partly on the contemporary view of education and partly on the social status of students. As regards the first aspect, teaching obviously meant more in a society which prized achievement than in a society in which one's social position depended almost entirely on one's birth. Educational qualifications replaced birth and personal favour as criteria for selection for employment. Higher education imparted additional skills of value in government, administration or various technical services. Education furthermore became a right of all men once it was thought that all men should have a voice in the choice of government. The provision of education thus became the duty of the state.

The other aspect which changed significantly was the relative social status of students and teachers. Before the Revolution, scientific subjects had been taught in the military academies[26] to officer cadets who were recruited almost exclusively from the nobility. Mathematical teachers like Monge and Lacroix were therefore instructing their social superiors. They themselves were neither officers nor did they belong to even the minor nobility. It was not until the Oath in the Tennis Court that the *Tiers Etat* asserted any political power. Lacroix,[27] teaching marine cadets under the *ancien régime*, contrasted the humiliation and insults he had formerly received from students and officers alike with the newly found independence and authority he enjoyed at the Ecole polytechnique, where students came from all social classes. Lacroix ended his career as dean of the Paris faculty of science.

Some German visitors[28] to France just after the revolutionary period were amazed to find among professors of science, government ministers and councillors of state, most notably Chaptal and Fourcroy. In the new France there was no lack of dignity accorded in the professorial function. To succeed such lecturers must have been the ambition of many a young man in their numerous audiences. It was as a teacher that a scientist was most exposed to the public, and the scientist was probably thought of as much as a lecturer as a research worker.

Before the Revolution the teaching of science[29] was usually conducted in one of three contexts. It was either at an elementary level, as in the schools and colleges, or at a popular level, as in the famous courses of natural philosophy of the abbé Nollet. It often reached its highest level in a military or naval context. Artillery or military engineering cadets in the 1770s and 80s received a very sound mathematical training, a training from which others might have benefited in a different system. The Revolution in creating a much wider view of scientific education was nevertheless able to draw on some of the traditions of the *ancien régime* and, most important, it had available a staff able to teach and examine; the staff included Monge, Lacroix and Laplace.

The Revolution marked both a quantitative and a qualitative change in the position of science. There was an increase in the scale of scientific activity, but there was also a change of pattern. Science not only had a significantly greater institutional support, it was now a nationally recognised activity inviting talented young men. The Revolution had replaced an aristocracy of birth by an aristocracy of achievement, or, as it has come to be called, a "meritocracy". Young men who, a generation earlier, would have followed in their fathers' footsteps, now thought out afresh their place in society. The tradition of respect in which science was held by the *philosophes* turned many in that direction. Science could now provide not only an intellectual challenge but a career. A young man who acquired a scientific education by availing himself of the free courses of lectures given in Paris could aspire to fame and fortune in this newly opened field. The establishment of a system of grants enabled students to benefit from a higher education without the traditional financial barriers. The Ecole polytechnique went further by providing not only grants but by inviting applications for places on a national basis, using the mechanism of a competitive examination to select the most talented. This institution therefore combined some of the best known men of science of the *ancien régime* on its staff with exceptionally able students. It was not long before some of the best of these students were themselves recruited on to the staff. The Ecole polytechnique thus played a dual role in the fostering of the profession of science. Not only did it provide the training by which Biot, Arago, Gay-Lussac, Poisson and others of their generation became scientists, but it provided posts for them afterwards.

Certification

Insofar as there was any systematic training at a high level in the physical sciences in the early years of the nineteenth century, it was provided largely by the Ecole polytechnique. As entrance to the Polytechnique depended on a competitive examination, it was in itself an achievement to have been accepted as a student. But to have undergone in addition an intensive course of study over two or three years was to have received one of the best scientific educations available at the time. The strong mathematical basis of the curriculum was supplemented by instruction in physics and chemistry, and there was a final examination. Unlike the military academies of the *ancien régime*, the science taught was not in a strictly vocational context. One of the aims of the Ecole polytechnique was to provide a broad scientific education.[30] Purely vocational training took place later in the écoles d'application, where one really learned the craft of the military or civil engineer, or some other technical skill which was taught in a school specially devoted to it.[31] To be a *polytechnicien* was not necessarily to be a scientist. Many graduates went into the army, became engineers or administrators. But, at least in the early years, many of the best became scientists. When they did so it was by virtue of their training, their publications and the posts to which they were appointed in recognised institutions. Their training as *polytechniciens* not only gave them a common social bond; it provided them with a recognised certification.

Certification was one of the primary functions of the university faculties established under Napoleon. The faculties of science, established in the years 1808–12 in Paris and various provincial centres,[32] suffered at first by being created later than the grandes écoles, having poorer facilities and lower prestige. The best faculty of science was that established in Paris and its posts were filled at a stroke of the pen by giving chairs as second appointments to established scientists who already held teaching positions in the grandes écoles. In principle, more posts had been created and hence there were more opportunities for young scientists to seek a professional base; but it took a generation for these opportunities to be realised.

In medicine the revolutionary period had witnessed a few years of anarchy with virtual freedom to practice without qualifications. A natural reaction came in the early years of the nineteenth century when the value of certified qualifications came to be doubly appreciated. This also affected science, which was also drawn into the higher educational system which was oriented towards examinations.

It had been an initial weakness of the faculties that they had been principally examining bodies with the power to award the grades of *bachelier, licencié, docteur* (and later *agrégé*)[33] in science. Yet in a society increasingly conscious of the value of formal qualification it became a positive advantage of the university system that it offered a series of grades which were useful and often necessary in gaining employment. Under the Restoration the *baccalauréat* became a basic qualification in any profession,[34] and the *licence* was a licence to teach required for all teachers except those in elementary schools. Crowning the University system was the doctorate by thesis, the official qualification of the faculty professor.[35] The first generation of French nineteenth-century scientists tended to be dominated by graduates of the Ecole polytechnique— Biot, Gay-Lussac, Arago, Poisson—but this famous school was often by-passed by the next generation, that of J. B. Dumas. It was with this generation that the doctorate really emerged as a professional qualification of the scientist.

In the 1830s doctoral candidates were not only more numerous than in the first twenty years of the faculties, but they were also more distinguished.[36] Among those submitting theses for the doctorate in the ten-year period 1832–41, we find the mathematicians Bertrand and Liouville, and the physicists Regnault, Despretz and A. E. Becquerel. Among those who later distinguished themselves in the biological sciences and mineralogy were Milne-Edwards, Quatrefages and Delafosse, while the chemists included such names as Dumas, Persoz, Boussingault, Pelouze, Peligot, Malaguti, Henri Sainte-Claire Deville, Laurent and Gerhardt. All of these did important scientific work and thirteen out of the seventeen mentioned became full members of the Académie des Sciences. It was to the benefit of French science that, in a country where the emphasis in the educational system was on the transmission of a culture, there was a specific incentive in the examination system to do research. It was only a pity that, instead of harvesting young men's early creative talents, the research was delayed until the final stage in the system when the candidate would usually be in his thirties or forties.

Research

Appointment to a teaching post usually meant not only a salary but the incidental facilities to do some research in a laboratory. Even though research might not be regarded as a part of the post, access to a laboratory, however simple, was a precious perquisite to many young teachers. It is right to emphasise the growth of science in the post-revolutionary era in the context of teaching, but there were other institutions which placed special emphasis on research. In addition to the Collège de France, the two institutions of particular importance were the Bureau des Longitudes and the Muséum d'histoire naturelle, dealing respectively with the physical sciences and the biological sciences.

The Bureau des longitudes[37] was much broader in its interests than its British counterpart. Admittedly it was the utilitarian aspect which was emphasised to the Convention when a case was being made for its establishment.[38] It was to be a great help to shipping and navigation. But the influential role of Laplace in the early years of the Bureau and the appointment of two mathematicians and four astronomers not only constituted a majority of the Bureau[39] but provided a nucleus of men with strong interests in the physical sciences, and they were able to pursue many of these interests at the Bureau. These salaried scientific civil servants had remarkable freedom for research within a broad framework. An inspection of the minutes of the Bureau reveals that they functioned rather like a scientific society, holding regular meetings at which not only astronomical data and navigational work, but also other pieces of research were presented and discussed.[40] In engaging officially on further research on the metric system—both the prolongation of the meridian and the comparison of standards—the Bureau was continuing work which had been initiated by the Académie des Sciences under the *ancien régime*.[41]

The Jardin du Roi had been a key centre for the study of natural history under the *ancien régime*, but when at the time of the Revolution it became the Muséum d'histoire naturelle and added significantly to its collections, it became something of a research centre, embracing chemistry, mineralogy, geology, botany, zoology and anatomy.[42] The staff of specialists not only had extensive collections on which to draw, but they were provided with houses within the grounds of the institution. They thus became a scientific community in a very special sense of the term, and their sense of community was helped by several other factors, notably the democratic administration. Instead of being under the control of an administrator appointed by the government, they all took a share in administration and were indeed referred to as *professeurs-administrateurs*. The weekly meetings of the professorial staff were held under the chairmanship of one of their number elected for one year only. The professors also published a journal, the *Annales du Muséum d'histoire naturelle*,[43] the existence of which greatly strengthened the research orientation of the Muséum, and there must have been informal pressure on the staff to contribute to it.

Stages of the Scientific Career

Anyone looking at the institutions organised in Paris or expanded after the Revolution is immediately impressed by the quality of the staff. The men who taught there are famous among persons with scientific training, and their students are now being studied.[44] However, the persons in the middle of the hierarchy, the senior students and the junior staff, who have received less attention, have also had a significant part in the establishment of science as a profession. Whether at the Ecole polytechnique and the Ecole normale, where their duties were largely teaching, or at the Muséum and the Bureau des Longitudes, where they were mainly research assistants, their posts have a fundamental significance in any assessment of the formation of the idea and structure of a scientific career. These young men received a salary and had the benefit of association with the most eminent scientists of France. In an establishment of higher education, they could deputise for the professor in his absence, thus acquiring valuable experience. Many aspiring scientists were called upon in this way. These assistants were paid to do scientific work and they were obvious candidates for any vacancy which might arise in the scientific hierarchy. At a school like the Ecole polytechnique, concerned with producing military and civil engineers and administrators as well as scientists, it was important for the future of science that some of the more able students should be retained at the school in a scientific atmosphere for further training so that their basic scientific education should not be "lost" to more immediate and practical concerns. The small salary which they received meant that they did not have to spend their time doing work unrelated to science in order to earn a livelihood.

Before the Revolution it was always theoretically possible for the bottle washer to succeed his master, but there was an immense gulf between the two. Such a succession would entail a very long period of close personal relationship. Thenard,[45] who began his scientific career as bottle washer to Fourcroy, was very much the exception in a generation which included scientists like Gay-Lussac, Biot and Arago, who all underwent formal courses of training. A major criticism of entry as a servant was that only occasionally did it develop into a system of apprenticeship. It was not an education but a substitute for one.[46]

To speak of "day-release" at this time might seem a terrible anachronism, yet if it were possible for a youth to obtain a general or scientific education at the same time as helping in a laboratory, he would have the basic requirements to become a scientist. Such an opportunity was provided by the Ecole polytechnique which appointed boys of fourteen and upwards as laboratory assistants (*aides laboratoires*) to help the students with their practical work.[47] The Ecole polytechnique was one of the first institutions to offer practical work in chemistry for students.[48] This called not only for teaching laboratories but also for staff. There were at least ten laboratory assistants but they were not required to be on duty all the time. It was intended that these young men should "find in this service a means of instruction"[49] and several hours each week were set aside for the study of mathematics. Although the boys were to be selected on grounds of intelligence, the position was in fact considered a valuable enough training by both deputies and members of the staff of the school for them to solicit such posts for young relatives or friends.[50] Unfortunately in 1797 a short-sighted effort at economy abolished the ten established posts of laboratory assistant.[51]

The creation of such posts was an enlightened step in the history of scientific education, but it could only provide basic instruction. What we are looking for is not how boys managed to benefit from the Ecole at a lower level than the entrance examination, but rather how it provided opportunities for post-graduate education for the better students at the end of their course. When in 1798 the ambitious original programme of a three-year course was cut to two years as a measure of economy,[52] the possibility of remaining a third year was left to several categories of students including "those who having done two years work wish to devote themselves to the study of a particular science of their choice . . ." and "students who, although not wishing to attach themselves to a public service,[53] wish to increase their knowledge of science and technology . . .". Because the extra year was a special privilege, students wishing to stay were required to take yet another competitive examination and were also liable to be called upon to help in the education of the more junior students.

From the foundation of the Ecole polytechnique there were to be section leaders or *chefs de brigade*, senior students who were rather like monitors as they were responsible for aiding the other students academically, keeping the register of attendance and helping generally to maintain discipline. At first there was the idea that these should be graduates of the school, but as there were no graduates in 1794 some of the senior and more able students—Malus and Biot among them—were given a concentrated course of preliminary training to enable them to become *chefs de brigade*, their final appointment depending on votes by the senior students themselves.[54] The monitorial system did not work well, particularly with regard to discipline.[55] Nor was it satisfactory academically, since the section leaders were spending so much of their time helping other students with their work instead of pursuing their own studies. It did provide, however, something of a stage in the scientific career, not only because of the duties of the post but because it provided an additional source of income.[56]

However, in 1798 the council of the Ecole polytechnique took a step which, although intended partly to contribute to the instruction of students, had a most important effect on the careers of the more able students of the school. This was the introduction of the post of *répétiteur*—sometimes translated inelegantly as "repeater"—to go over the work in mathematics with the students.[57] These were to be annual appointments given, as far as possible, to former section-leaders who wished to take up teaching as a career. One of the first of the two such appointments was given to Francoeur, one of the original students of 1794 who eventually became a mathematics professor at one of the Paris lycées. It was also decided to change the permanent post of demonstrator—*préparateur*—in chemistry to a similar annual appointment of two assistant demonstrators, preferably former students who were interested in practical chemistry. They recruited Desormes—a student of 1794—and also Thenard, who had not had the advantage of being a student there.

The post of *préparateur* for chemistry was soon re-named répétiteur, and a post was created for physics, too.[58] Thus, mathematics, physics, and chemistry each had a junior post with an annual salary of 1500 francs. Although the appointments were intended to be annual, the appointees were re-elected each year until 1804 when other vacancies arose.[59] The two *répétiteurs* for mathematics moved on to the more senior position of entrance examiners, Thenard was appointed to the chair of chemistry at the Collège de France, and Desormes resigned to set up in business as a chemical manufacturer. The vacancies so caused gave a first chance to other bright young men. We thus find Ampère, formerly holding a lycée post at Lyon, given the opportunity of a post in the capital. Gay-Lussac, who had already been ear-marked for a future vacancy with the title of répétiteur-adjoint, became a full répétiteur for chemistry. His duties were of two main kinds. As the title indicated, he was expected to help students both with difficulties in lectures and in practical work. But he was also the demonstrator for lectures, setting up apparatus and acting as store-keeper. He thus acquired good practical experience in an academic environment and when Fourcroy, the professor of chemistry, died in 1809, Gay-Lussac seemed the obvious successor to the post. Other professors in the first half of the nineteenth century at the Ecole polytechnique, who began as répétiteurs, include the mathematicians and physicists Poisson, Cauchy, Mathieu, Liouville, Savary and Regnault. Of course promotion was by no means automatic. Auguste Comte was able to obtain the post of répétiteur but not a higher one, despite his many solicitations. Le Verrier was promoted not within the school but to the Observatoire, of which he became, in time, a most distinguished director.

By the 1830s the répétiteur was generally recognised as a grade in the scientific career. An American visitor who made a study of the French educational system at this time remarked: "The utility of this latter class of teachers is well established in France, and they are found in every institution in which lecturing is practiced. . . ."[60] He went on to explain that this system had a double advantage. As it was clearly understood that the professors only had to deliver formal lectures—and not actually to teach students—eminent scientists were willing to accept chairs. The student was the richer both in consequence of hearing lectures by experts and also by being questioned on the substance of the lectures by the *répétiteur*. The *répétiteurs* also benefited. It was reported that "young men of talent seek the situation of répétiteurs as the best method of showing their particular qualifications, and the most certain road to a professorship".

Of the many careers followed by graduates of the Ecole polytechnique, one which offered only a few opportunities of entry but which was high in quality was the Bureau des Longitudes. In the course of a discussion with Cassini in the opening years of the nineteenth century, Delambre made the following observations, of some importance in our study of the structure of careers in French science

> "Is not the Ecole Polytechnique a forcing ground (*pépinière*), where one will always find students talented in mathematics and physics to fill the positions of *adjoint* [at the Bureau des Longitudes]? But in spite of the confidence which must be inspired by the ability of these students, they are still required to justify their admissions. They enter the Observatoire as assistants. There they make observations and calculations, and when their talent is assured, they proceed to the rank of *adjoint*, which gives them a living. The *adjoints* have the chance of becoming full members [of the *Bureau*] which finally determines their future, and provides them with double their salary, as well as the right of being chairmen at the meetings of the Bureau and of voting on all its decisions."[61]

We find that the post of astronomer at the Bureau des Longitudes—a senior post with the considerable salary of 6000 francs—was consistently filled from the ranks of the assistant astronomers. Lefrancais Lalande, Bouvard, Burckhardt, Biot and Arago, who were the first five young scientists to hold posts as assistants, were all in turn promoted to the post of astronomer.

At the Muséum d'histoire naturelle, the post of assistant naturalist with a curatorial function was created very early in its new post-revolutionary organisation. In June 1794 the Committee of Public Safety agreed to establish, in addition to the professorial posts, two additional posts described as *naturalistes conservateurs*.[62] Their duties were related particularly to the arrangement and conservation of specimens. They were to be responsible to the professors, and they were to have a salary about half that of a professor. When this suggestion was approved, the professors asked for three posts at a slightly lower salary. The Committee of Public Safety[63] agreed, and the title of assistant—*aide naturaliste*—began to be used for these three posts, soon to be increased to four.

It might have been possible to have increased the number of assistants gradually without any great deliberation about their precise function. However, certain events in 1803 were to lead not only to an increase in the number of such assistants but to an appraisal of their value and in particular their place in the structure of the career in the natural sciences in France. It began when professors of chemistry Fourcroy and A-L Brongniart, at a meeting on 1 June 1803, made a case for an additional junior post for chemical analysis. This would be a new post of "assistant chemist" (*aide-chimiste*) comparable to the "assistant naturalist", but he would be "attached to the professors of chemistry and responsible in particular for the analysis of different natural history specimens".[64] It was agreed that special funds should be asked for this post. However, this brought the whole question of assistants under discussion and at what must have been a lively meeting it was agreed to establish a committee[65] to report on the question of number, duties and salaries of assistants. The number required might depend not only on the development of a particular science but the collections in that department and the needs of teaching.

The report of this committee is an important document in the development of the profession of science in France since it analyses in detail the professional role of assistants in the Museum. It first pointed out that the title of *aide-naturaliste* had hitherto been given to young men of quite different educational background and function. There were those whom we would call scientists and a second group whom we should call laboratory assistants or technicians. In the words of the report "The first group are true naturalists, whose work is almost entirely scientific; the others are technicians (*des artistes*) whose talent, although undoubtedly precious, does not however presuppose any formal education." The Committee recommended that only the former group should be called assistant naturalists. It proposed the creation of two additional positions in this category.[66] The report continued

> "Together with the three existing posts and with that which has just been created for chemistry, these two assistants would form a kind of intermediate body between the professors and the other employees, an honourable training ground (*une pépinière*) where excellent naturalists will be produced and where [future] professors can with advantage be recruited."

With the acceptance of this report by their colleagues, the designation "assistant naturalist" was no longer a miscellaneous description among the lesser employees, gardeners and warders of the Muséum; it had become a rank. From then on staff lists of the Muséum gave first the professors, secondly the *aides naturalistes*, and thirdly the miscellaneous staff required by the Muséum. The post of assistant naturalist was to have a tenure of five years, but in compensation for its temporary nature it had now a definite status, a position in the hierarchy calling on definite skills and providing valuable experience. In the early nineteenth century we find among their ranks Valenciennes, Laugier, Audouin[67] and Isidore Geoffroy Saint-Hilaire,[68] all future professors at the Muséum where they had been trained. Promotions were by no means automatic but for the naturalists mentioned and others it was a definite grade in the career of those who followed the new profession of science in nineteenth-century France.

A fourth institution which cannot be omitted in any discussion, however brief, of science in nineteenth-century France is the Ecole normale, which in the Restoration became the leading institution to prepare for the competitive examination of agrégation; this was the qualification of the more highly paid teachers. As the qualifications for lycée and university posts were raised, what was needed in science now was not so much an inducement to keep a graduate from the Ecole polytechnique or a graduate with a licence in a suitable atmosphere of scientific research but to do this for someone with the higher degree of *agrégation*. This meant the creation of another research position, that of the *agrégé préparateur*, which might be translated into English as graduate demonstrator. At the Ecole normale, Pasteur became the first agrégé préparateur (1846–48)[69] and the position of demonstrator was henceforth reserved for graduates of the school who stayed on to prepare their doctoral theses in association and consultation with the professors of the school.

Pasteur later wrote in enthusiastic terms to the Minister of Education

"I know by experience how much the leisure of these modest positions is worth for a young man who had been touched by the fire of science. [He benefits from] an atmosphere of healthy study amongst well-endowed laboratories, under the benevolent direction of proven masters."[70]

The position of *agrégé préparateur* was really an extension of the post of *répétiteur* of the Ecole polytechnique in the early nineteenth century. Both had responsibilities for helping more junior students with practical work and thus justified their small salaries in the eyes of the Ministry of Education. But they were also in immediate contact with the professors who could guide them with their research. The post of *agrégé préparateur* raised the demonstrator to a new and higher level. It kept him in higher education and provided conditions which enabled him to obtain the qualification to become a university scientist for life.

In the 1830s, Dumas proposed that there should be a system for the attachment of agrégés to the faculty of science.[71] It was about time he felt that the faculties took responsibility for guiding research for the doctorate. Positive steps in the establishment of a sequence of stages of a career in French science were taken in the 1840s with the creation of the post of *agrégé-préparateur* at the Ecole normale, in the 1850s with encouragement for senior students at the Ecole normale to stay on and do their doctoral research, and finally in the 1860s, with the establishment of the Ecole pratique des hautes études.[72] The Ecole pratique is sometimes seen as atypical of the French scientific scene—a belated move in 1868 after half a century of neglect. Professor Ben-David calls it "the first [French] experiment in postgraduate training".[73] I believe that a closer study will show it merely as a logical conclusion and consolidation of a system which had been implicit in the French higher educational system since the very beginning of the nineteenth century. Unfortunately it was not sufficiently general; it depended on private as well as state laboratories, and until the Second Empire it was not part of any national plan. The main weakness in the structure of this career ladder was that whereas the *grandes écoles* were able to attract the best students, they were not connected with the network of faculties of science with their many teaching posts and the obvious possibility of a career within them.[74] Only the Ecole normale cut across and joined these two systems. The Ecole normale thus became increasingly important in the opportunities for a scientific career in France; it enabled French scientists, at the same time, to benefit from the prestige and conditions of a *grande école* and to obtain the qualification necessary for employment in the faculties.

Britain and France

The professionalisation of science in Britain came later than in France. In a sentence which has often been quoted, Charles Babbage in 1851 could write that "science in England is not a profession; its cultivators are scarcely recognised even as a class".[75] I have tried to show here how, half a century earlier in France, science had already become a profession. It was a profession with its own specialised journals, so much so that a French writer in the 1820s could complain that in his country "the literary journals seldom concern themselves with the sciences".[76] This situation contrasts with Britain, where periodicals such as the *Edinburgh Review* and the *Quarterly Review* gave serious treatment to science in a society where science was still a part of the culture of an educated middle-class rather than a specialised activity offering full-time employment and remuneration. Much reputable science in early nineteenth-century Britain was done in the setting of literary and philosophical societies,[77] where science was still formally associated with literature as polite culture rather than a specialised intellectual activity. It was said of the early nineteenth-century English geologist William Smith that "geology had kept him poor all his life by consuming his professional gains";[78] for other British men of science, one could, instead of "geology", write "chemistry" or "astronomy" or just "the pursuit of science". After John Dalton had published his atomic theory, he continued in nineteenth-century Britain to give the sort of tuition in elementary science and mathematics which reminds one of the situation in France under the *ancien régime*.

This almost menial employment of men of science contrasts with the situation in France after 1800. David Brewster gave the following account of employment of British men of science

> "Some of them squeeze out a miserable sustenance as teachers of elementary mathematics in our military academies, where they submit to mortification not easily borne by an enlightened mind. More waste their hours in drudgery of private lecturing, while not a few are torn from the fascination of original research and compelled to waste their strength in the composition of treatises for periodical works and popular compilations."[79]

Thus in some ways Britain in the 1830s invites comparison with France in, perhaps, 1780.

Of course Babbage and Brewster tended to exaggerate the lack of recognition of science in Britain. If there was a cultural lag, there were also important cultural differences, and the British tradition of individualism, local initiative, and independence from government control conferred benefits in other ways. When, eventually, science became a profession in Britain, it had a French model not necessarily to copy, but to adapt to the contemporary state of science, British national traditions and the different social conditions of Victorian Britain.

References

1. For analyses of the profession of science, see J. Ben-David (1972), "The Profession of Science and its Powers", *Minerva*, **x**, 362–83; and E. Shils (1968), "The Profession of Science", *Advancement of Science*, **xxiv**, 469–80. This paper was published in *Minerva* (1975), **xiii**, 38–57. I should like to thank Edward Shils for various editorial suggestions on its presentation.
2. See R. Hahn (1971), *The Anatomy of a Scientific Institution: The Paris Academy of Sciences, 1666–1803* (Berkeley, University of California Press).
3. In 1794 it was merely stated that candidates should have a knowledge of arithmetic and the elements of algebra and geometry, but already in 1795 the syllabus of the entrance examination included trigonometry, conic sections and the solution of algebraic equations up to the fourth power. A. Fourcy (1828), *Histoire de l'Ecole Polytechnique* (Paris), 30, 82.
4. In the period of the Directory the external examiners were Laplace and Bossut. The Danish astronomer Thomas Bugge attended some of these examinations in 1798 and said that Laplace was likely to fail several candidates that year, which meant that they could not proceed to the higher schools of specialised vocational training. *Science in France in the Revolutionary Era, described by Thomas Bugge* (ed. M. P. Crosland, Cambridge, Mass., Massachusetts Institute of Technology Press, 1969), 42. Laplace undoubtedly had a powerful influence on the content and standard of the courses.

F

5. J. B. Morrell (1971), "Individualism and the Structure of British Science in 1830", *Historical Studies in the Physical Sciences*, **iii**, 183–204.
6. Count Leclerc de Buffon, *De la manière d'étudier et traiter l'histoire naturelle*.
7. A standard source for the part played by science and scientists in the war effort is: C. Richard (1922), *Le Comité de Salut Public et les Fabrications de Guerre* (Paris). See also C. C. Gillispie (1957), "The Natural History of Industry", *Isis*, **xlviii**, 398–407. For the role of artists in political propaganda, see D. L. Dowd (June 1948), *Pageant Master of the Republic: Jacques Louis David and the French Revolution* (Lincoln, Nebraska, University of Nebraska Studies), New Series, no. 3.
8. Condorcet strongly advocated the teaching of science as a means of eradicating superstition.
9. For an account of the patronage of science in Napoleonic France, see M. P. Crosland (1967), *The Society of Arcueil: A View of French Science at the Time of Napoleon I* (London, Heinemann Educational Books), chapter 1, pp. 4–55.
10. The *Journal des Savants*, which published book reviews and general articles covering the whole field of learning, began publication in 1665, two months before the *Philosophical Transactions of the Royal Society*.
11. *Annales de mathématiques pures et appliquées*, 21 vols., 1810–31 (ed. Gergonne).
12. *Annales de chimie*, 96 vols., 1789–1815. Its distinguished editorial board was originally headed by Lavoisier and included most leading French chemists. In 1816 the journal became the *Annales de chimie et de physique* under the joint editorship of Gay-Lussac and Arago, and for at least another half century it was the leading French journal in both chemistry and physics.
13. *Annales du Muséum d'Histoire Naturelle*, the "house journal" of the Museum.
14. See H. Guerlac (1970), "Lavoisier", in *Dictionary of Scientific Biography* (ed. C. C. Gillispie, New York, Scribners), **viii**, 67–91.
15. Letter from Laplace to Lagrange, 11 February 1784, in *Oeuvres de Lagrange* (ed. J. A. Serret), **xiv**, 1892, 130.
16. Caritat de Condorcet, *Oeuvres* (eds. A. Condorcet O'Connor and D. F. J. Arago, Paris, 1847), **vii**, 423.
17. "Rapport fait au Conseil des 500 par Villers au nom de la Commission des Dépenses, dans la Séance du 2 Prairial au 4." Quoted in L. Aucoc (1889), *L'Institut de France* (Paris), 38.
18. The category of *Academicien libre* was also abolished. The reintroduction of this category by Louis XVIII in 1816 was merely a manifestation of his blindness to what had happened in France in the Revolutionary and Napoleonic period.
19. Such a situation is, of course, quite different from that of the wealthy amateur, who is free to spend as much time as he wishes on science and can suddenly decide to employ his leisure in some other way.
20. "Rapport et projet de décret sur l'établissement d'une Ecole centrale de santé à Paris", *Réimpression de l'Ancien Moniteur*, **xxii** (Paris, 1794), p. 665. Although the reference of this passage was to medical education (hence "les sciences utiles" above), Fourcroy was also concerned about the same time with the establishment of the Ecole polytechnique.
21. See E. Bezout (1770–2), *Cours de mathématiques à l'usage du corps royal de l'artillerie*, 4 vols. (Paris). The bibliography of Bezout's textbooks is complex and is further complicated by the fact that, even after his death, further textbooks for military and naval cadets were published based on his work.
22. The lectures were published as *Séances des Ecoles Normales, recueillies par sténographes et revus par les professeurs*, 2nd ed., 10 vols. (Paris, 1800, 1801). A further three volumes of *Débats* were also published.

23. It lasted only three months but was reconstituted by Napoleon in 1808.

24. A full biography of Gay-Lussac by the present writer is in preparation.

25. Gay-Lussac and Biot asked in 1815 for permission to divide the course, the former giving lectures on gases, heat, etc., while the latter assumed responsibility for lectures on optics, sound and magnetism (Archives nationales, F17, 1933).

26. See *Enseignement et Diffusion des sciences en France au XVIIIe siècle* (ed. R. Taton, Paris, Hermann, 1964).

27. S. F. Lacroix (1805), *Essai sur l'enseignement en général* (Paris), 128n.

28. J. F. Reichardt (1896), *Un hiver à Paris sous le Consulat, 1802–1803* (Paris), 456.

29. See R. Taton, ref. 26.

30. C. A. Prieur, one of the founders of the Ecole polytechnique, had insisted that the school should produce not only engineers but also architects, industrialists, science teachers and simply "enlightened citizens". "Mémoire sur l'Ecole Centrale des Travaux Publics" (Messidor an 3), *Procès-verbaux du Comité d'Instruction Publique*, ed. Guillaume, **vi**, 302.

31. The principal écoles d'application (also called écoles de services publics) included the Ecole des ponts et chaussées, Ecole des mines, Ecole des géographes, Ecole de génie, Ecole d'artillerie and Ecole des ingénieurs de vaisseaux.

32. Faculties of science were established at Paris, Aix, Caen, Dijon, Lyon, Strasbourg and Toulouse. By 1812 faculties were established at Bescançon, Grenoble, Metz and Montpellier. Further faculties of science were established in towns which soon reverted to Italy, Switzerland and Belgium; these are, therefore, not listed. For another assessment of the faculties in the Restoration period, see R. Fox (1973), "Scientific Enterprise and the Patronage of Research in France, 1800–1870", *Minerva*, **xi**, 442–73.

33. The *agrégation* was a very difficult competitive examination taken after the *license* but before the *doctorat*. Although the origin of the *agrégation* (candidates were "agrégé" or attached, to the university) goes back to the eighteenth century, it was only in 1821 that an examination was held. There were then three sections: grammar, letters and science. In 1840 the *agrégation* in science was divided into two, with candidates choosing between mathematics and experimental sciences. By 1842, 58 per cent of teachers and administrators in state secondary schools held the *agrégation*. A. Prost (1968), *Histoire de l'Enseignement en France, 1800–1907* (Paris, Armand Colin), 72–3.

34. The *Ordonnance* of 13 September 1820 stated that the *baccalauréat* was intended "ouvrir l'entrée à toutes les professions civiles et devenir pour la société une garantie essentielle de la capacité de ceux qu'elle admettrait à la servir".

35. Two theses were in fact required. Article 24 of the decree of 17 March 1808 on the establishment of the faculties states that for the doctorate "on soutiendra deux thèses, soit sur la mécanique et l'astronomie, soit sur la physique et la chimie, soit sur les trois parties de l'histoire naturelle, suivant celle de ces sciences à laquelle on déclare se destiner".

36. A. Maire (1892), *Catalogue des Thèses de Sciences soutenus en France de 1810 à 1890* (Paris). I should like to thank John Christie, who drew the existence of this catalogue to my attention.

37. A brief account of the establishment of the French Bureau des Longitudes is given in M. P. Crosland, ref. 9, 209–13.

38. *Le Moniteur*, **xxv**, 84–7 (11 Messidor an 3 = 29 June 1795).

39. The staff of the original *Bureau* of 1795 was as follows: mathematicians—Lagrange, Laplace; astronomers—Lalande, Cassini, Méchain, Delambre; "anciens naviga-teurs"—Borda, Bougainville; geographer—Buache; technician—Carochez.

40. Examples of pure scientific rescarch carried out under the auspices of the Bureau des Longitudes include work in optics on the polarization of light and interference, also on the velocity of sound.

41. C. Bigourdan (1901), *Le système métrique des poids et mesures* (Paris, Gauthier-Villars).

42. M. Deleuze (1823), *Histoire et description du Muséum d'Histoire Naturelle*, 2 vols. (Paris).

43. First series, 20 vols., 1802–1813. Continued as *Mémoires du Muséum d'Histoire Naturelle*, 20 vols., 1815–1832.

44. M. Bradley (1974), "The Ecole Polytechnique: 1795–1830—Organisational Changes and Students". M.Phil. dissertation, University of Leeds.

45. P. Thenard (1950), *Un grand Français: le Chimiste Thenard, 1777–1857, par son fils; avec introduction et notes de Georges Bouchard*, Dijon. Although Thenard began under the old apprenticeship system, he was able to take advantage of the new organisation of science inasmuch as he was appointed *répétiteur* at the Ecole polytechnique in December 1798, no doubt through the good offices of Fourcroy. Thenard thus represents a transitional case in the recruitment of scientists.

46. For an example in the Napoleonic period of a laboratory assistant who was encouraged to take university examinations and become a professional scientist, see the case of J. E. Bérard (1789–1869) in M. P. Crosland, ref. 9, p. 134.

47. A. Fourcy (1928), *Histoire de l'Ecole Polytechnique* (Paris), 56–7.

48. W. A. Smeaton (1954), "The Early History of Laboratory Instruction in Chemistry at the Ecole Polytechnique, Paris, and elsewhere", *Annals of Science*, **x**, 224–33.

49. Organization de l'Ecole Polytechnique, 30 ventose an 4, Titre IV, Art. XVIII, *Journal de l'Ecole Polytechnique*, cahier 3, Prairial an 4.

50. Ref. 47, p. 92.

51. Ref. 47, p. 109.

52. Ref. 47, p. 137.

53. For example the School of Bridges and Highways.

54. The first *chefs de brigade* were taken from former students of the Ecole des ponts et chaussées, the Ecole des mines, or on the basis of their marks in the entry exam. In 1794, the 50 senior students were called *aspirans-instructeurs*; it was the intention to divide them into two groups, the most mature becoming *chefs de brigade* immediately while the remainder might become *chefs de brigade* later. It was Monge who decided that the 50 students themselves should vote for the *chefs de brigade* (Ref. 47, pp. 60–1, 71n).

55. Ref. 47, pp. 161, 241–2.

56. Under the Directory, students usually received 360–500 francs per annum. For section leaders this increased to 700 francs (M. Crosland, ref. 4, p. 201).

57. Ref. 47, pp. 159–60. Under the *ancien régime* the use of advanced students to help in teaching had been introduced at the Ecole du corps royal du génie at Mézières and at the Ecole des ponts et chaussées. See F. B. Artz (1966), *The Development of Technical Educations in France, 1500–1850* (Cambridge, Mass., Massachusetts Institute of Technology Press), 84, 100.

58. Ref. 47, p. 257.

59. Ref. 47, pp. 272–3.

60. A. D. Bache (1839), *Report on Education in France* (Philadelphia), 568–9.

61. M. G. Bigourdan (1928), "Le Bureau des Longitudes: Son histoire et ses travaux de l'origine (1795) à ce jour". *Annuaire du Bureau des Longitudes*, A39. Quoted by J. A. Cawood (1974), *The Scientific Work of D. F. J. Arago (1786–1853)*. Ph.D. thesis, University of Leeds, 8–9.

62. *Recueil des Actes du Comité de Salut Public* (1901) (ed. F. A. Aulard), **xxiv**, 153 (17 Prairial an 2–5 June 1794). Under the *ancien régime* at the Jardin des Plantes, the position of demonstrator had been recognised in addition to that of professor, but this was primarily (as in Renaissance faculties of medicine) a distinction of function between practical and theoretical.

63. *Extrait des Registres du Comité de Salut Public*, 26 Prairial an 2, discussed at a meeting of the Museum professors on 28 Prairial—Archives nationales AJ XV, 577.

64. Archives nationales AJ XV 590, Séance du 12 Prairial an XI, Lettre au ministre, 15 Prairial an XI.

65. At the meeting of 19 Prairial (8 June 1803) it was agreed that the committee should consist of Lacépède, Lamarck, Geoffroy Saint-Hilaire and Cuvier—all naturalists. However, the committee which reported also included the chemist Fourcroy. The committee was thus an impressively large one, consisting of most of the leading figures at the Museum.

66. With responsibilities for quadrupeds and birds, reptiles and worms, respectively.

67. Jean-Victor Audouin (1797–1841) was one of the founders of the Société d'histoire naturelle de Paris in 1822. In 1830 he replaced Latreille as Assistant naturalist at the Museum and, on Latreille's death three years later, he succeeded him as professor of zoology.

68. Isidore Geoffroy Saint-Hilaire (1805–61), the only son of his more famous father Etienne, was engaged as *aide naturaliste* at the age of 19. He succeeded to his father's chair at the Museum in 1841.

69. R. Dulou and A. Kirrmann (Sept. 1973), "Le laboratoire de chimie de l'Ecole Normale Superieure", *Bulletin de la Société des amis de l'Ecole Normale Superieure*, No. Hors serie, p. 8.

70. L. Pasteur to the Minister of Public Instruction, 24 October 1858, *Pasteur, Correspondence* (ed. P. Vallery Radot, Paris, Flammarion, 1940–51), **ii**, 37.

71. "Extrait des procès-verbaux des deliberations de la Faculté des Sciences", Séance du 15 Novembre 1837, O. Gréard (1889), *Education et Instruction, Enseignement Superieur*, 2nd ed. (Paris), 242, 248.

72. V. Duruy, *L'Administration de l'Instruction Publique de 1863 à 1869* (Paris, n.d. (1878?)), 644–58.

73. J. Ben-David (1971), *The Scientist's Role in Society: A Comparative Study* (Englewood Cliffs, New Jersey), 103.

74. R. Gilpin (1968), *France in the Age of the Scientific State* (Princeton, New Jersey, Princeton University Press), 112.

75. C. Babbage (1851), *The Exposition of 1851* (London), 189.

76. Baron de Ferussac (1822), *Bulletin général et universel des annonces et des nouvelles scientifiques*, **i**, Prospectus, p. 1.

77. One of the most prominent of these societies was the Manchester Literary and Philosophical Society, where John Dalton found encouragement and in whose *Memoirs* he published much important work.

78. W. Walker (1862), *Memoirs of Distinguished Men of Science of Great Britain, Living in 1807–1808* (London), quoted by D. S. L. Cardwell (1972), *The Organization of Science in England*, 2nd ed. (London, Heinemann Educational Books), 17.

79. *Quarterly Review* (1830), **xliii**, 327.

10. German Science in the Romantic Period*

D. M. KNIGHT
(*University of Durham*)

The first edition of Kant's *Critique of Pure Reason* appeared in 1781, and Hegel died in 1831; and these dates form the approximate limits of our period.[1] In a different light, the period is the active life of Goethe, whose *Werther* came out in 1774 and who died in 1832; for as well as Kant's "Copernican Revolution" in philosophy, we must take note of the great movements in German literature, the *Sturm und Drang* of Goethe and Schiller and the Romanticism of the next generation. But the period also includes most of the work of Werner, the geologist, and Gauss, the mathematician, and abroad, of William Herschel and Alexander von Humboldt. While the connexions of science with philosophy and literature were close in Germany at this time, there were strong autonomous traditions in medicine, in mathematics, in astronomy, and in chemistry and mineralogy, with their close association with mining. The student of German science in these years must not be too concerned with those whose approach to nature seems very different from ours, or from that of the later nineteenth century; we must try to keep the whole spectrum of science in view. And in Germany as elsewhere, the dictum that all science is either physics or stamp-collecting should be dismissed by the historian interested in the early nineteenth century; for then chemistry and the biological and geological sciences seemed much more important and exciting than mechanics.

If part of our problem is then to decide what constituted *science* in our period, another and equally intractable part is to decide what constituted *Germany*. Whereas France and Britain had their great metropolitan centres, Paris, London and Edinburgh, Germany had no capital city and no clear frontiers. Men of science brought up in German states often made their reputations abroad, particularly in Russia, and in Britain where the links with Hanover made migration easy in the first part of our period. If we take *Germany* to mean those parts of Europe where German was the language spoken—and by our

* The contributor would like to thank Dr Barry Gower for helpful remarks made on an early draft of this paper.

period, generally also written even for learned and scientific works—then we find ourselves contemplating an area not unlike the Germany of 1939; but in our period this was composed of numerous states, some very small, and each with its court, which were in various degrees friendly or hostile to one another. Science as carried on in Prussia, in Bavaria and in Austria had much in common, but it was not being carried on in one country; we must never forget the diversity of Germany when we compare what was happening there with what was going on in England or France.

This paper is intended to be no more than a review of some of the work that has been going on in this field. For, although I have been interested for a number of years in ideas emanating from Germany around the year 1800, I would not claim to have a very specialised knowledge of the history of science in Germany. Indeed, like some well known English men of science of the nineteenth century, I have begun to learn German several times.

A national tradition in science may be characterised by distinct intellectual assumptions or paradigms, by distinct institutions or degree of professional consciousness among men of science, or by distinct relations of theory and practice. To an outsider, it looks as though German science in our period differed from science as carried on in France or Britain in all these ways. In the realm of institutions, it was from Germany that two of the most important features of nineteenth-century science took their origin; the teaching laboratory, which is discussed by Wilfred Farrar, and the Association for the Advancement of Science. Both these became features of science as carried on in all countries in the Victorian period. Scientists emerged as a group, and science perhaps as a profession, when they came to receive a training which was different from that given to those studying arts subjects; and this separateness and group feeling was encouraged by the annual meetings of the various Associations for the Advancement of Science, where all those engaged in any way with the various sciences could meet each other and keep up to date, and could hope to make their collective voice heard.

Both these institutions were developed in Germany at about the same time. It is curious that while Liebig, whose teaching laboratory was the first of a long line of them, is reverently remembered in chemical textbooks and brief histories of science; the name of Lorenz Oken, who was responsible for calling together the *Naturforscher* in the meetings that formed a model for other Associations for the Advancement of Science, is almost forgotten. Both institutions were equally important and successful, and this difference must be due to the very different intellectual traditions in which the two men worked; this may indicate that one cannot separate the institutional history of science from the intellectual, to which it is in the last resort subservient.

Chemistry in the 1820s was emerging from a period of revolutionary change, with exciting but rather vague theories, into a phase more resembling "normal science" where the need was for more information to fill in a picture in which the outlines seemed to have been already sketched in. This was a field in which questions were difficult to answer rather than difficult to ask, and Liebig's teaching prepared students to set about answering the relatively straightforward questions posed in the chemistry of his day. Chemistry was no longer a matter of "chemical philosophy", closely involved with a complete worldview, as it had been in Germany since the days of Paracelsus; nor was it simply a branch or ally of medicine; it was an empirical science, in which any reasonably able and hardworking student could expect to make a contribution to knowledge.[2] Chemistry for most students became a workaday affair.

Liebig was in reaction to the very different conception of science in general and chemistry in particular which had been current in his youth, and which he had shaken off in France. This was the world-view to which Oken subscribed, and like Liebig he ran his own journal to propagate it, although the meetings of the *Naturforscher* do not seem to have been used by him or others to disseminate a particular view of science. Oken belonged to the school of thought, if this is not too precise a term, which was called *Naturphilosophie*, and which seems to have no exact parallels in other countries at the time.

Naturphilosophie consisted of very broad and general views about the world. It can be traced back to the Platonic belief in the rationality of the world, as A. O. Lovejoy indicates in the last chapter of his *Great Chain of Being*; and it has strong connexions too with the "chemical philosophy" of Paracelsus and van Helmont. But the immediate source of *Naturphilosophie* was Schelling, who strongly influenced some important Romantic men of science—most notably, Ritter and Oersted.[3] In *Naturphilosophie*, an important idea is that because the human mind is a part of nature and reflects nature, the laws of our reason must resemble those of nature, and we could therefore hope to make scientific discoveries from an armchair. This is really a version of the old belief in the analogy between the microcosm, man, and the macrocosm, the great world of nature. But it seemed more up-to-date because Kant had argued that the categories into which we organise knowledge are arrived at independently of experience; the idea that empirical knowledge could be attained *a priori* was not extraordinary in Schelling's Germany.

This did not mean that devotees of *Naturphilosophie* did not bother to perform experiments. Cartesians with clear and distinct ideas had nevertheless tested their deductions by experiment, and Ritter and Oersted were keen experimenters and discovered a number of very interesting phenomena. In Oersted's miscellaneous writings, translated into English as *The Soul in Nature* in 1852,[4] we find encomia both on the human reason and on experiments. But it did mean that experiments played a less important part in the view of science held and taught by adherents of *Naturphilosophie* than in that of Liebig or other adherents of more "normal" science. This belief in the power of the reason was not shared even by Romantic scientists in other countries; Davy for example often urged his hearers to beware of persuasive theories that lacked experimental support, and while men of science in England often stressed the value of the imagination in science they saw it merely as suggesting analogies and not as a guide to the truth.

In *Naturphilosophie*, the great generalisation which the reason supplied was that force rather than matter was the underlying reality which men of science must investigate. In place of the materialism associated with eighteenth-century Rationalism and Enlightenment, in which everything—even thought —was accounted for in principle in terms of matter and motion, Schelling and his disciples saw a world of forces. This was a world of flux, in which solid objects or persons only endured like waterfalls or columns of smoke do, through the perpetual flux of their material particles. Discoveries in physiology, a science in which Germans were prominent in the eighteenth century, reinforced this view. The material particles indeed were unimportant or even imaginary; our hand seems to encounter solid particles when we press it on the table, but really as Oersted remarked we only encounter forces.

These forces must be indestructible, as matter had been for the materialists, though they will manifest themselves in different ways, as electricity, magnetism, heat or chemical affinity. This rather general prediction seemed to be verified in discoveries made in the last years of the eighteenth century and the first years of the nineteenth. Galvani established a relationship between electricity and muscular motion, and the work of Volta, Davy and Berzelius showed the connexion of electricity and chemical affinity. With his discovery of infrared radiation, the German expatriate William Herschel showed the close connexion between light and heat; and Ritter then discovered ultraviolet radiation, with its capacity to induce chemical reactions, and thus connected light with chemical affinity. Finally, Oersted discovered electromagnetism, which was a connexion for which he had long argued and sought.

These connexions marked a step towards the principle of conservation of energy,[5] and made the old view that light, heat, and so on were the effects of distinct imponderable fluids much less plausible. But none of these studies were quantitative, and we shall be misled if we concentrate simply upon the use made of them a generation or two later on. What seems to have struck contemporaries was that these connexions proved the primacy of force, and the interconvertability of the various forces of nature; as Davy put it, electricity and chemical affinity were manifestations of one "power".[6]

This was clearly a step towards the unity, simplicity and harmony of nature which was predicted in the Platonic tradition; in the world of the Romantic scientist, there is nothing cold and inanimate. We seem to encounter brute matter, but really we meet living forces. It appeared that chemists and physiologists had stormed the bastille of materialism in which Rationalists had allowed their minds to be imprisoned. Moreover these were sciences in which the Germans traditionally excelled, whereas materialism was viewed by

Romantics in Germany and in England as a particularly French kind of intellectual pox. Whereas Rationalists had seemed to give the impression that everything could already be explained in principle, to Romantics there now opened up a vast but misty vision of a much more organic universe. The prospects for science became indefinite.

It was not only the primacy and unity of force which reason seemed to dictate, and which experiments duly revealed; for the forces also seemed to display a polar quality. Thus electricity is negative or positive, and magnetic poles are north or south seeking; and chemical affinities showed themselves in a tendency to combine either with oxygen, or with hydrogen or metals. Indeed it was in accounting for chemical reactions that the conception of polar forces was more successful than anything that had gone before. Unlike gravitation, chemical attraction was elective—some things will react together and others will not. If, as the work of Davy and Berzelius seemed to establish and as Schelling and his disciples seemed to have predicted, different substances have different characteristic electrical states which are neutralised in reaction, then this elective character is to be expected. Two positively charged substances will not unite, without the mediation at least of a negatively charged one. Davy showed that an electric charge would modify the properties of things; that positively-charged silver becomes chemically reactive, for example, and negatively-charged zinc inert.

Polar opposites were not simply at enmity; for their relationship was dialectical.[7] Negatively-charged oxygen and positively-charged hydrogen do not annihilate each other but come together in a new synthesis yielding water. This was the pattern to be found throughout the realm of chemistry, where the emergence of new qualities in the course of chemical reaction had always been perplexing. To those who thought like this of polarities and dialectical syntheses, any kind of atomic theory was anathema. Not only was nature for them, as it had been for Leibniz, characterised by continuity, so that the sharp boundary between an atom and the void seemed intolerable, but also atomic theories seemed not to provide any genuine explanation. How could the juxtaposition of inert and massy particles generate new qualities? To followers of *Naturphilosophie*, atomic theories seemed, as they had earlier to the phlogistonist Stahl, "to scratch the surface of things and leave the kernel untouched".[8]

Chemistry as pursued by devotees of *Naturphilosophie* was thus a dynamical science, and Lavoisier's quantification of science in terms of weights seemed irrelevant. This was not far from the mainstream of chemical thought in the eighteenth century, for phlogiston in some contexts does seem to have represented something similar to what we would call chemical energy, while attempts to quantify affinities were a feature of late-eighteenth-century chemistry in most countries.[9] And indeed we find that dynamical accounts of phenomena in early nineteenth century chemistry were sometimes labelled "phlogistic"; certainly some theories of electrochemistry which seem rooted in the phlogiston theory, with its rejection of the decomposition of water, are in fact very close to "dynamical" ones with their stress upon the forces which modify matter. Where Davy's thoughts most resembled Ritter's, he labelled his conjectures "phlogistic".[10] There are certainly aspects of *Naturphilosophie* which are old-fashioned rather than wild or visionary, and we must be careful about approaching it backwards, from the nineteenth or twentieth centuries, because this must distort our vision of it.

This continuity with the chemistry of the eighteenth century, and with the older "chemical philosophy" which had always insisted that chemistry must be dynamical and that it was more fundamental than mechanics, had a certain basis in nationalism. This was a relatively new phenomenon in Germany, and was brought into prominence by the wars with France in our period. We find the same phenomenon in England, where a good hit at a French theory could be expected to bring applause. But the greatest men of science were above these kind of rivalries for at least some of the time, and nostalgia, for a chemistry as it used to be before the French innovations, was not the sole motive which turned men's minds towards *Naturphilosophie*.

More important was the close link in Germany at this time between science and philosophy. There may have been two cultures in Napoleonic France,[11] but it seems safe to say that there were not in contemporary Germany. In the Universities, theology had by the latter years of the eighteenth century been dethroned as Queen of the Sciences. Her place had been taken by philosophy, and the sciences were thus as rule taught in the philosophy faculty. They were not therefore taught as "normal science", but as genuine natural philosophy involved with a world-view and abounding in generalisations. Those learning the sciences had to learn more than a number of techniques; and conversely, those learning philosophy had to learn some science. This was evident in their writings later on. Like any other process in history, the separation of science from other activities brought disadvantages as well as advantages.

It is perhaps because of this close connexion between men of science and men of letters that scholars have found German science in our period fascinating. Certainly it is the philosophers and philosopher-scientists who have *recently* received most attention, and the time has now come to look more closely at what experts have been writing about them. We must throughout this discussion bear in mind that it may not have been these men who were considered of major importance in the scientific world—particularly outside Germany—in their own day. We will begin with Goethe,[12] for upon Goethe's science we have recently had a little book by H. B. Nisbet, as well as more specialised articles and a reprint of the English version of the *Theory of Colours*.

Nisbet's study is particularly interesting because he emphasises the extent to which Goethe's science was typical of that of his day, and was rooted in various living traditions. It was not therefore a matter of "wayward but intriguing theories" unique to Goethe, whose originality, in Nisbet's view, consisted chiefly in the way he related his ideas to give a comprehensive theory of life. Goethe was characterised by a belief in the unity, continuity and harmony of nature, coming ultimately from the Platonic tradition, by a Baconian devotion to experiments and descriptive science, and by a rejection of the Rationalists' devotion to mechanical or mathematical explanations. Science for Goethe must not be abstract; in any situation, one must search for the *Urphänomen*, which is a genuine phenomenon but has an ideal character because it is particularly striking and has in some way the character of an archetype. An *Urphänomen* cannot be explained any further; when confronted by one, such as the different colours seen in engraved glass figures when they are viewed in different lights, the proper reaction is wonder. To try to go further would be infantile, like looking for one's twin behind a mirror.

In Goethe's science, man was not separated from nature, either as a dispassionate observer or as the framer of some abstract and highly theoretical explanation of the appearances. Rationalists who behaved like this were, in Goethe's view, studying abstractions or constructions of their own minds and not nature at all. The great authority for this kind of scholasticism seemed to be Newton, whose name stood in Germany as in England as a symbol of applied mathematicians generally. But whereas in England Newton's authority among men of science could not be challenged, Germans could not forget his quarrel with Leibniz; so that, whereas in England to be scornful of Newton would be an antiscientific gesture, in Germany it would indicate only opposition to the overweening claims of applied mathematicians and their publicists.

Goethe did contribute to anatomy, and his refusal to separate man from nature or to allow jumps in nature lay behind his recognition of the inter-maxillary bone in man. Similarly his *Theory of Colours* is an important work in psychological optics; his recognition that depth can be given to a picture by colour, and that classical perspective does not adequately account for the way we perceive sizes and distances, were of some interest to artists—and it is not surprising that the English translation (of the less-polemical parts) of the *Theory of Colours* was by Charles Eastlake, who was both a Royal Academician and a Fellow of the Royal Society. Goethe perceived the dangers of a narrow and technical science, concerned with questions which seem to be chosen because they are answerable, but which are either of no general interest or are completely abstract and bogus, scholastic in the pejorative sense. Goethe's own life showed this refusal to specialise and narrow down his interests; among his associates were some experts or "walking encyclopedias" to whom he turned for information, contenting himself with principles. He paid the price of not becoming an expert, in that he did not make "really significant contributions" to the progress of science; but on the other hand he did incorporate his science into an attractive world-view.

Colours for Goethe were the outcome of the opposition of light and dark, and he believed that an *Urphänomen* would often have a polar character. Goethe claimed to have got the idea of polarity from Kant's discussion of attraction and repulsion, with its conclusion that these were essential properties of matter. But polarity played a larger part in Goethe's thinking than it had in Kant's, and was to play an even larger part in the thinking of Schelling and his disciples. Goethe was not a Romantic; he belonged to an older generation, and although Romantics derived inspiration from his works he considered them over-subjective—for he believed that his science was truly objective, unlike that of either the Newtonians or the followers of *Naturphilosophie*.

Because of his attempts to lay down the forms and limits of empirical science *a priori*, Schelling has received some attention from philosophers of science.[13] However, he has received more attention in recent years from historians, because of his influence—real or supposed—upon the development of electro-chemistry, electromagnetism and energetics. To be interested in *Naturphilosophie* because it contributed to the main stream of modern science is to do what is pejoratively called "whig history"; perhaps it can be better described as "applied history of science", of which the "pure" branch is concerned with studying past science for its own sake, and seeing it in the context of its own day. If we try to do applied history before we have done pure history, we shall get a distorted vision of the past; and we shall be all too prone to invoke mysterious "influences" connecting apparently similar theories held at widely different times or places. Such connexions are all too often like the hypothetical bridges which late Victorian Darwinians invoked to connect distant continents in the remote past, which constituted an obstacle to a more economical way of theorising in geology.[14]

In a long article published in 1973, Barry Gower has reviewed recent work on *Naturphilosophie* in the light of his extensive knowledge of the printed and manuscript sources in German. He is sceptical about the impact of Schelling's philosophy on any definite scientific problem, for it seems to him so general as not to lead to any definite predictions in science, but rather to "a preoccupation with metaphysical issues of a very broad and abstract nature". He analyses in some detail the work of Ritter and of Oersted, which did have general connexions with Schelling's metaphysical scheme. But he does not find direct and obvious links between the two, which is what we might expect, since many of those who discovered connexions between forces were not disciples of Schelling, and conversely the discoveries of Ritter and Oersted were incorporated into the science of the day, which outside Germany did not become much tinged with *Naturphilosophie*.

Gower's sceptical conclusion is therefore that in this episode in the history of the relations of science and philosophy, we should look less for empirical predictions made *a priori* and more for scientists responding to certain philosophical problems raised in the context of scientific explanation. This view is particularly welcome to the historian of science because, while, in Gower's exposition, such men of science as Ritter and Oersted appear as unusually interested in philosophy of science and indeed metaphysically-minded, they do not appear as wild and visionary persons engaged upon some barely comprehensible enterprise, which nevertheless led in an ill-defined way to the concept of energy. Historians should not be put off Gower's paper by its length and stiffness, for it contains grist for their mill.

It seems then that Schelling and his disciples were more concerned with explanation than with prediction, and the same seems true of Hegel.[15] Two translations of his *Philosophy of Nature* have recently been published; Michael Petry's is fully-annotated, and provides one of the best routes into German science in the early nineteenth century that we have. Hegel was not a Romantic, and much of his philosophy was directed against that of Schelling; that his science had much in common with Schelling's is a further indication that we must beware of attributing too much directly to *Naturphilosophie*. Hegel emerges in Petry's study as remarkably well-informed in the sciences; more so than we might expect of a philosopher even when science was chiefly taught in the philosophy faculty. Petry emphasises, however, that his interest was chiefly in established science, in what was taken for granted and taught in curricula, rather than in discovery and in problems on the frontiers of natural knowledge.

Goethe, Romantic authors and chemists had all protested against the programme, conveniently associated with the name of Newton, of reducing all natural science to Mechanics. Hegel designed a structure, a classification of the sciences, in which reduction would be seen to be impossible. Hegel's science is a matter of levels, with mechanism at the bottom, the chemical sciences above it, and above them the life sciences. Any attempt to explain the phenomena of one science in terms of a science on a different level was doomed to failure. Thus chemistry could explain the cause of death in a particular case, but it could never give an account of the cause of life, for the biological sciences are on a different and higher level. Hegel was well aware of the unity of nature, as for example Schelling and Goethe were; but he also saw its diversity, and he was reluctant to fall in with what to him seemed the loose way of thinking which grouped electricity, galvanism and chemical affinity as different aspects of one phenomenon. He saw them as related, but as being on different levels; as being indeed three adjacent steps in the stair-case of knowledge. To group them all together would lead only to confusion and vagueness. The small steps or transitions were carefully worked out by Hegel, and we nominalists, who take it for granted that the frontiers between the sciences are matters of convenience alone, can with profit ponder the Hegelian scheme in which the boundaries are natural. As Petry points out, Hegel made some mistakes in applying his own scheme—notably in opposing Newton's physical optics to Goethe's psychological optics which he could have put on different levels and reconciled, and he made some few blunders also. But in general, his account of the sciences is reliable, and his *Philosophy of Nature* shows the value of the German system in which philosophers knew enough science to provide a well-informed general picture of it.

The phenomena of chemistry and electricity were particularly puzzling in our period, and in their polar aspects they displayed analogies with human love and hate which made them interesting to philosophers.[16] In these sciences, investigators were likely to turn philosophical, or to turn to philosophers, as they worked towards acceptable explanations and nomenclature. Physiology too attracted the philosophically-minded; Goethe and Oken made their contributions to this science, stressing the unity of the organism and relying upon analogies which to some contemporaries and most successors appeared far-fetched. But there were holistic physiologists in Britain, France and Germany who were much less metaphysically-minded than Oken. In such fields as natural history, geology and mathematics, men of science in Germany seem to have been no more involved with philosophy than were their contemporary colleagues in other lands. The strength of the physiological tradition doubtless meant that organic analogies were taken further in the sciences in Germany than they were elsewhere. But we must not allow ourselves to be dazzled by all virtuosity of German metaphysicians of our period, and ascribe to them any vitalistic, holistic or dynamical emphases in German science. Englishmen who looked across the North Sea to Germany in the early years of the nineteenth century did not find *Naturphilosophie* filling their field of view, and nor should we.

We have already touched upon the difficult and often unprofitable question of the possible influence of *Naturphilosophie* on the development of electrochemistry, electromagnetism and energetics. Neither Davy nor Faraday, whose ideas show some affinities with some of those of Ritter and Oersted, read German; and such papers of Ritter's as did appear in English journals in the early nineteenth century were translations from the French versions and were highly abbreviated, though this was not stated.[17] That is, the facts discovered were reported, but not the reasoning behind their discovery. As we noted above, Gower indicates that there was a resemblance between the theories of Ritter thus presented and the phlogiston theory still being advocated by Joseph Priestley, and by lesser-known men such as Gibbes, whose papers appeared in the same journals as the English accounts of Ritter's work. In fact Ritter's theory was not really very like that of Gibbes, as Gower demonstrates. It depended upon an abstruse analysis of what took place at interfaces between solid and liquid or between different metals; but in an abstract which had passed through French such recondite speculation had been lost.

In Davy's circle, Beddoes and Coleridge did know German, and possibly Davy could have got an accurate knowledge of Ritter's thought from them, though both of them were very creative thinkers, and unlikely to transmit a speculative theory without modifying it. Davy does refer in his "phlogistic" conjectures to the work of Ritter, as a support for his own Priestleyan view that water was not decomposed in electrolysis, but that oxygen and hydrogen were compounds of water and electricity. But even here Davy's and Ritter's views were not the same; in general it seems implausible to suppose that Davy would have found Ritter's thought congenial, for he was much more empirically-minded. It seems more plausible to suggest that he would be pleased to know that Ritter's views fitted in, as far as he cared to pursue them, with a dynamical view of chemistry. And an Englishman who was a correspondent of Priestley, as Davy was, would not need a spur from Germany to turn him towards a world view in which force was predominant, and chemistry and electricity the keys to a real understanding of the mutations of matter. Davy was in touch with a dynamical Newtonian tradition;[18] he was also interested in alchemical speculations and he had some knowledge of what Ritter was doing, but we must not put too much weight upon this last connexion, and make Schelling the godfather of potassium.

We should note that although Davy introduced polarity into chemistry in his Bakerian Lecture to the Royal Society in 1806, he does not use the word "polarity" or its derivatives. There seems no need to invoke an influence from Germany to account for Davy's bringing together chemistry and electricity, for this was an enterprise upon which chemists all over Europe had been engaged ever since Volta described his cell some six years before Davy's lecture. One did not need to be a devotee of Schelling's metaphysical scheme in order to appreciate the value of an electrical account of chemical affinity, though to anyone imbued with *Naturphilosophie* Davy's work would no doubt have been particularly gratifying.

It was from France that outstanding recognition came for Davy's lecture, for he was awarded the prize established by the *Institut* for making the greatest advance in electrical science in the year. A pillar of the *Institut*, and one who had himself done work in electrochemistry, was Georges Cuvier; in his *Rapport Historique* of 1810 he devoted some space, in his discussion of chemistry since 1789, to the work of Winterl,[19] who was associated with *Naturphilosophie*. The account is unsympathetic, but must indicate at least that Cuvier had read some speculative chemistry coming from Germany. It also indicates that he saw little real resemblance between the work of Davy and that of Winterl, whose theories he presents as seductive, but not based upon experience. One would not expect that Cuvier's report would have won disciples for *Naturphilosophie* in France or England, but it may have done so, just as Lyell's criticisms of Lamarck seem to have brought his writings out of obscurity.

There was already in English a German dynamical textbook of chemistry, F. C. Gren's *Principles of Modern Chemistry*, which was published in 1800.[20] But here again it is doubtful whether readers of this book would have got a clear

G

idea of the kind of chemistry associated with *Naturphilosophie*. Like most chemistry books of the time, it is a compendium of information, and the theoretical part does not bear a very close relation to the bulk of the text, and does not display the extravagant generalisation and speculation that we find in, for example, Oken's *Principles of Physiophilosophy*, which was not translated until 1847, when it was hardly a contribution to modern science.

It seems most plausible, therefore, to regard *Naturphilosophie* as a phenomenon largely confined to Germany. It is not less interesting if it is seen as local in time and space; after all, it is distinct national traditions that form the subject of this symposium. Nor does it involve any "cutting down to size"; we shall appreciate the work of Oersted and Ritter better when we see it in its own proper context, and such study will cast more light on the problems facing chemists and electricians in the early nineteenth century than will further conjectures about influences. Gower's attempt to determine the real connexion, or lack of it, between the metaphysical scheme of Schelling to which they adhered and the actual discoveries that were made by Ritter, Oersted and others—and they did make important discoveries—marks a more fruitful way of proceeding. The historian should surely not concern himself only or chiefly with episodes leading in a direct line to what is taught in science courses today. That is not the way history goes, though such an approach may provide a useful schema for those teaching the sciences. Gower is surely right in insisting that we shall see *Naturphilosophie* more clearly if we look at it rather than at its supposed effects.

We have suggested that while *Naturphilosophie* was a uniquely German scientific phenomenon, it was not the whole of German science. It belongs to the intellectual side of science, though it was reinforced by the institutional arrangement which put the sciences in the philosophy faculty. The best broad treatment in English of science in Germany in our period is still that of J. T. Merz; there we can begin, though naturally some of our interests and emphases will be different from his.[21] He refers to the small part which the Universities played in the history of the natural sciences in Germany in the later years of the eighteenth century, the great exception being Göttingen when Haller taught there. Merz dates a new state of affairs from the very beginning of the nineteenth century, when Gauss was appointed to a chair at Göttingen and soon showed himself the equal of the great French mathematicians. But it was not, in Merz's view, until about the end of our period that mathematical and experimental science began to play a prominent part in the German University system generally.

Merz does remark in the German Universities in our period the appearance of "the ideal of pure science and its pursuit". He does not seem to mean by "pure science" what we might mean, that is something set against "applied science"; but rather he seeks to convey the term *Wissenschaft*, an ideal of exact knowledge firmly based upon evidence whether in physics or history, which will altogether fit into a pattern. It does certainly seem to be true that even in our period the German Universities were devoted to research and scholarship to a higher degree than those elsewhere. There seems to be no reason to suppose that a distinction like that made in the later nineteenth century between pure and applied science was made in Germany in our period; all science was expected to be useful sooner or later, and *Wissenschaft* was not a term used in opposition to "useful knowledge".

Even *Naturphilosophie* had its utilitarian aspects, in its close connexions with physiology and thus with medicine as exemplified in the work of Oken, whom we have already met as founder of the annual assemblies of *Naturforscher*. Because Germany had no metropolis, these meetings were held in a different city each year, and thus provided a model for similar meetings in other countries where metropolitan arrogance was resented by those living in the provinces, or which like the USA had no real metropolis either.[22] Germany had, as we have remarked, a great tradition in physiology and medicine, associated with the names of Haller just before our period and of Blumenbach during it.[23] Because therefore Germany was at this time a poor country with a good educational system, physicians and natural historians trained in Germany often found work beyond its boundaries, where their professional competence opened doors to them.

Thus George and Reinhold Forster sailed on Cook's second voyage, in which he circumnavigated Antarctica and proved there was no habitable Terra Australis Incognita; they were the natural historians appointed when the arrangements made with Joseph Banks fell through.[24] The African Association in which Banks played a prominent role, employed John Lewis Burckhardt to explore Nubia and Arabia, in disguise; earlier they had sent Freidrich Conrad Hornemann to Africa.[25] Later the British Government employed Heinrich Barth to complete the exploration of the Niger, begun in our period by Mungo Park, Hugh Clapperton and the Landers. The expedition of 1817 under Captain Tuckey, which was sponsored by the Royal Society, had a Copenhagen-trained naturalist aboard, though one must record that he could not preserve, describe, or draw marine creatures seen through a microscope. And beyond the end of our period, the Prussian draft-dodger Ludwig Leichhardt was one of the most remarkable explorers of the interior of Australia. William Beckford, "England's wealthiest son", took a German physician with him on his tour of Portugal in 1794. This Dr Erhardt seems to have been rather solemn and devoted to his profession; but many of those trained in medicine were keen natural historians, and we must remember that at this time—when the distinction between pure and applied science was hardly made anywhere[26]—natural history seemed a particularly useful and practical discipline, having connexions with medicine, agriculture, textiles and mining. Banks' correspondence is a useful lead to Germans working on natural history in England, or under British auspices.

Abraham Gottfried Werner at Freiberg, where he lectured at the Mining Academy for forty years from 1775, established a great school of mineralogists; his disciples occupied an important place in geology throughout our period, when it was a science of great economic importance—as Rudolph Glauber had noted a century earlier in his essay on the Prosperity of Germany.[27] In Russia as in Britain, German men of science could find employment; the great naturalist P. S. Pallas held a position of enormous importance as adviser to the Empress Catherine the Great on the various expeditions sent out to explore Siberia and Russian America at the end of the eighteenth century. But the paragon of natural historians and travellers was Alexander von Humboldt, who not only observed but generalised. Like his eminent contemporary Robert Brown, Humboldt was less interested in non-descripts than in the whole aspect of the fauna and flora of a region. Humboldt seems to have been unusual in that his foreign connexions were with France; it was under French auspices that he was able to travel to Latin America, and it was in Paris that his researches were written up and published. In later years, Spix and Martius on the Amazon, and Richard and Robert Schomburgk in Guiana, filled in outlines that their countryman Humboldt had drawn. His brother William von Humboldt, founder of the University of Berlin, with his studies on language both transformed linguistics and changed the emphasis in anthropology.

In astronomy,[28] the work of Euler and Meyer made possible the method of determining longitude by measuring the distance apart of the Sun and Moon, providing a solution to a practical problem at the same period that Harrison's chronometer was perfected. Both methods were used, to find their position, by seamen and travellers throughout our period. Gauss computed the orbit of the minor planet Ceres, and his statistics enabled astronomers to average out a number of observations in a more exact and less intuitive manner. Fraunhofer's reputation as an optician stood very high, so that long after his death his telescopes were in great repute; Durham University for example bought one in 1839. Among the leading observational astronomers were J. F. Bessel, who first detected stellar parallax, H. W. M. Olbers, and Friedrich Struve, who went to Russia and worked first at Dorpat and then at Pulkowa. Others were William and John Herschel; William had come from Germany, and the Herschels kept up German connexions and German customs.

There were therefore many regions in which German science, whether done at home or abroad, was not high-flown, metaphysical and idiosyncratic, though the problems to which Germans addressed themselves even here were no doubt somewhat different from those which Englishmen or Frenchmen saw. Perhaps because many of them worked abroad, and those at home worked under various governments, but perhaps also because such men as Blumenbach, Gauss, Humboldt and Struve were great generalisers, Germans were great promoters of international co-operation in science. This was particularly successful in astronomical and geodetic observations, where Struve and Humboldt stressed the value of contemporary and comparable measurements of terrestrial magnetism and other phenomena at widely-separated places on the Earth. Such measurements soon, in the words of the geologist and traveller Strzelecki: "rendered the detached, unconnected, and minor observations of occasional observers of little or no value". The plans and objectives of Humboldt were sufficiently practical and practicable to recommend themselves to governments, who did send out well equipped expeditions to swing pendulums and magnets in remote places.[29]

The stolid and indefatigable German, carefully observing and painstakingly classifying or reducing observations, was as much a feature of our period as the philosopher imposing a pattern upon the facts of chemistry or physiology. From nineteenth-century science as from Nature, "we receive but what we give", and we should not limit our vision with blinkers.[30] In looking at science in Germany at this time, as in looking at fauna and flora, we must not be completely diverted from the norm, and from the whole aspect, by a few extraordinary productions. We must not be so dazzled by the change in chemistry between Ritter and Liebig as to see the whole of German science as chiefly a matter of metaphysics down to the 1820s and of empiricism thereafter; for this is a caricature which will make it more difficult for us to understand the frequently-close relations between German scientists and their contemporaries in other countries.

References

1. See H. G. Schenk (1966), *The Mind of the European Romantics* (London), and on the period in general, see my review-article in *History of Science*, **ix** (1970), 54–75, and W. D. Wetzels (1971), "Aspects of Natural Science in German Romanticism", *Studies in Romanticism*, **x**, 44–59.

2. J. B. Morrell (1972), "The Chemist Breeders: the research schools of Liebig and Thomas Thomson", *Ambix*, **xix**, 1–46, and the paper by W. H. Brock on Liebig's laboratory accounts which immediately follows it.

3. F. W. J. Schelling (1966), *On University Studies* (tr. E. S. Morgan, ed. N. Guterman; Athens, Ohio); B. S. Gower (1973), "Speculation in Physics; the History and Practice of *Naturphilosophie*", *Studies in the History and Philosophy of Science*, **iii**, 301–56; and see H. A. M. Snelders' (1970/1) papers on Winterl and on Schweigger in *Isis*, **lxi**, 231–40 and **lxii**, 328–38. J. W. Ritter (1968), *Die Begründung der Elektrochemie* (ed. A. Hermann, Ostwalds Klassiker, Frankfurt am Main). See also H. C. Oersted (1852), *The Soul in Nature* (trs. L. and J. B. Horner, London); and L. P. Williams (1966), *The Origins of Field Theory* (New York), ch. 2, and *Michael Faraday* (1965), London, pp. 137ff.

4. For a comparison of Oersted and Davy, see my paper (1967) "The Scientist as Sage", *Studies in Romanticism*, **vi**, 65–88.

5. T. S. Kuhn (1959), "Energy Conservation as an Example of Simultaneous Discovery", in *Critical Problems in the History of Science* (ed. M. Clagett, Madison, Wis.), 321–56. P. F. Dahl (1972), *Ludvig Colding and the Conservation of Energy Principle*, New York.

6. *Collected Works of Sir H. Davy*, 9 vols. (ed. J. Davy, London, 1839–40), **viii**, 284.

7. S. T. Coleridge (1969), *The Friend* (ed. B. Rooke, London), **i**, 94, 479 (where the footnote should refer to Ritter rather than to "Volta and La Place"), 493–4; *Hints towards the Formation of a more comprehensive Theory of Life* (ed. S. B. Watson, London, 1848).

8. Quoted by J. R. Partington (1961), *A History of Chemistry*, **ii** (London), 665.

9. A. M. Duncan (1970), "The Functions of Affinity Tables and Lavoisier's List of Elements", *Ambix*, **xvii**, 28–42, and his introduction to the reprint of T. Bergman (1970), *Dissertation on Elective Attractions* (1785), (London), and see H. Davy (1807), *Philosophical Transactions*, **xcvii**, 42.

10. R. Siegfried (1964), "The Phlogistic Conjectures of Humphry Davy", *Chymia*, **ix**, 117–24; J. Davy (1836), *Memoirs of the Life of Sir H. Davy*, 2 vols., (London), **i**, 40eff.

11. See Maurice Crosland on the French university faculties, p. 141.

12. H. B. Nisbet (1972), *Goethe and the Scientific Tradition*, (London); G. A. Wells (1966–7), "Goethe and the Intermaxillary Bone", *British Journal for the History of Science*, **iii**, 348–61; Goethe (1966), *Conversations and Encounters* (eds. and trs. D. Luke and R. Pick, London); Goethe (1967), *Theory of Colours* (tr. C. L. Eastlake, 1840 reprint, London). See also Nisbet (1970), *Herder and the Philosophy and History of Science*, (Cambridge); and A. Gillies (1969), *A Hebridean in Goethe's Weimar*, (Oxford).

13. On *a priori* science, see R. Harré (1965), *The Anticipation of Nature*, (London); on Naturphilosophie, see Gower (ref. 3).

14. A. Hallam (1973), *A Revolution in the Earth Sciences*, (London).

15. G. W. F. Hegel, *Philosophy of Nature* (tr. A. V. Miller, London, 1970; and ed. and tr. M. J. Petry, 3 vols., London, 1971). See also W. Kaufmann (1966), *Hegel*, (London).

16. See my paper (1972), "Chemistry, Physiology, and Materialism in the Romantic Period", *Durham University Journal*, **lxiv**, 139–45; T. McFarland (1969), *Coleridge and the Pantheist Tradition*, (Oxford); T. H. Levere (1971), *Affinity and Matter*, (Oxford); H. W. Piper (1962), *The Active Universe*, (London).

17. There is some discussion of scientific journals in my *Sources for the History of Science, 1660–1914*, London, 1975. On Franco-German relations in chemistry in the early part of our period, see M. P. Crosland (1962), *Historical Studies in the Language of Chemistry*, (London), p. 207ff.

18. Davy was later indignant that any credit for "his" electrochemical theory should be given to the Germans; J. Davy, *Life of Sir H. Davy*, **i**, pp. 326–7—but of course many people dislike paying intellectual debts. On Davy and phlogiston, see ref. 10. On dynamical Newtonianism, A. Thackray (1970), *Atoms and Powers*, Cambridge, Mass.; J. E. McGuire (1968), "Force, Active Principles, and Newton's Invisible Realm", *Ambix*, **xv**, 154–208; R. E. Schofield (1970), *Mechanism and Materialism*, (Princeton); P. M. Heiman (1973), "Nature is a perpetual worker", *Ambix*, **xx**, 1–25; M. A. Sutton (1971), "J. F. Daniell and the Boscovichean Atom", *Studies in History and Philosophy of Science*, **i**, 277–92.

19. G. Cuvier (1810), *Rapport Historique sur les Progrès des Sciences Naturelles depuis 1789*, (Paris), pp. 83–8; some of Winterl's work is described in F. Szabadváry (1966), *History of Analytical Chemistry* (tr. G. Svehla, Oxford), pp. 48–9.

20. F. C. Gren (1800), *Principles of Modern Chemistry*, 2 vols., (London); there is a discussion of chemistry books in my *Natural Science Books in English*, (London, 1973), chap. 7. L. Oken (1847), *Elements of Physiophilosophy* (tr. A. Tulk, London).

21. J. T. Merz (1896–1914), *A History of European Thought in the Nineteenth Century*, 4 vols., (Edinburgh); the reference is to vol. **i**, p. 211.

22. A. D. Orange (1972), "Origins of the British Association", *British Journal for the History of Science*, **vi**, 152–76. *Victorian Science* (eds. G. Basalla, W. Coleman and R. H. Kargon, New York, 1970). See also two studies of the British scientific community in this period, by J. B. Morrell (1971), in *Historical Studies in the Physical Sciences*, **iii**, 183–204; and by S. Shapin and A. Thackray (1974), *History of Science*, **xii**, 1–28.

23. L. S. King (1966), introduction to A. Haller, *First Lines of Physiology, 1786*, (reprint New York); and his *Road to Medical Enlightenment*, London, 1970.

24. J. Cook (1955–74), *Journals* (ed. J. C. Beaglehole, 4 vols. in 5, Cambridge), 1955–74, vol. **ii**; *The Banks Letters* (ed. W. Dawson, London, 1958), and supplement, *Bulletin of the British Museum (Natural History)*, *Historical Series*, **iii** (1962–9), 71–93; A. M. Lysaght (1971), *Joseph Banks in Newfoundland and Labrador*, (London)—this work offers a reinterpretation of Banks' work generally; and see H. B. Carter (1974), "Sir Joseph Banks and the Plant Collection sent from Kew to Russia", *Bulletin of the British Museum* (Natural History), *Historical Series*, **iv**, 281–385. The Forsters' bird-paintings have recently been printed; G. Steiner and L. Baege (1971), *Vogel der Südsee*, (Leipzig); and some of their fishes are included in P. J. P. Whitehead (1968), *Forty Drawings of Fishes made by the Artists who accompanied Captain James Cook*, (London).

25. On exploration from Britain in this period, see my paper "Science and Professionalism in England", *Proceedings of the XIV International Congress on the History of Science*, Tokyo, 1974. Ferdinand Bauer accompanied Robert Brown on Flinders' survey of Australia; a sumptuous edition of his flower paintings made then has been announced, London. See also A. M. Coates (1973), *The Book of Flowers*, (London); and P. Mitchell (1973), *European Flower Painters*, (London). E. W. Bovill (1968), *The Golden Trade of the Moors*, 2nd edn., (London), and *The Niger Explored* (1968, London); *Barth's Travels in Nigeria* (ed. A. H. M. Kirk-Greene, London, 1962); J. K. Tuckey (1818), *The River Zaire*, (London), pp. lxiiff, 49—despite Oersted, Copenhagen perhaps hardly counts as "German" even in the loose sense of our period; *The Letters of F. W. L. Leichhardt* (ed. and tr. M. Aurousseau, 3 vols., Cambridge, 1968); W. Beckford, *An Excursion to the Monasteries of Alcobaça and Batalha* (ed. B. Alexander, Fontwell, 1972).

26. A. P. Molella and N. Reingold (1973), "Theorists and Ingenious Mechanics", *Science Studies*, **iii**, 323–51; M. Berman (1972), "The Early Years of the Royal Institution", *Science Studies*, **ii**, 205–40.

27. R. Glauber (1689), *Works*, (London); M. Sauer (1802), *Expedition to the Northern Parts of Russia*, reprint Richmond, 1972; A. Chamisso sailed as naturalist on Kotzebue's circumnavigation in 1815–18, *To the Pacific and Arctic with Beechey* (ed. B. M. Gough, Cambridge, 1973), p. 149. Humboldt's *Essai sur la Géographie des Plantes*, Paris, 1805, really 1807, has been reprinted, London, 1959, by the Society for the Bibliography of Natural History; and a reprint of his *Voyage Aux Regions Equinoxiales du Nouveau Continent*, 30 vols., Paris, 1805–34, has been published, Amsterdam and New York, 1971–3. See L. Kellner (1963), *Alexander von Humboldt*, (London). J. B. Spix and C. F. P. Martius (1824), *Travels in Brazil in the years 1817–1820*, 2 vols., (London). Humboldt was a pupil of the great generaliser and founder of physical anthropology, J. F. Blumenbach, whose *Anthropological Treatises* were translated into English by T. Bendyshe and published in London, 1865. Flourens, on p. 52 of this, compares Blumenbach favourably as a follower of the positive and soundest methods with Oken, a devotee of *systems*. But such enterprises as Humboldt's *Cosmos*, "A Sketch of a Physical Description of the Universe" (English tr. by E. C. Otté *et al.*, 5 vols., London, 1849–58) would not have been

undertaken by anybody except a German during our period. Some of Humboldt's interest is that he combined attention to detail with a capacity for the broadest generalisation—and for the latter at least his German training can be given the credit.

28. *Tobias Mayer's Opera Inedita* (ed. and tr. E. Forbes, London, 1971), and *The Euler-Mayer Correspondence*, 1751–5, London, 1971. T. Hall (1970), *Carl Friedrich Gauss*, (Cambridge, Mass.). C. W. Dunnington (1955), *Carl Friedrich Gauss*, (New York). H. Woolf (1959), *The Transits of Venus*, (Princeton, N.J.), p. 189ff. A. M. Clerke (1885), *History of Astronomy during the Nineteenth Century*, (Edinburgh); R. Grant, *History of Physical Astronomy*, 1852, reprint, New York, 1966; *Durham University Calendar*, 1840, pp. xiff, 18. On the Herschels' Christmas tree, J. W. Clarke and T. M. Hughes (1890), *Life of Adam Sedgwick*, (Cambridge), **ii**, 101.

29. P. E. Strzelecki (1845), *Physical Description of New South Wales and Van Diemen's Land*, (London), p. 48. W. Cannon (1964), "History in Depth", *History of Science*, **iii**, 20–38. *The Admiralty Manual of Scientific Enquiry* (ed. J. F. W. Herschel, London, 1849); reprint, introduced by me, London, 1974. *Herschel at the Cape* (eds. D. S. Evans *et al.*, Austin, Texas, 1969).

30. The quotation is from Coleridge's Dejection Ode, stanza IV, and paraphrases a passage in Plotinus.

11. Science and the German University System, 1790-1850

W. V. FARRAR

(University of Manchester Institute of Science and Technology)

One thinks of science nowadays as essentially an activity pursued in universities. Virtually every working scientist has been taught in one, and—despite the growth and importance of industrial and government research—the majority of published papers come from universities or related institutions.[1] It was not always so. From the Renaissance until well into the nineteenth century, the connexion between science and universities was a casual one. Of those who made significant contributions to science during this long period, some were educated in universities, many were not. Science was advanced by clergymen, schoolmasters, engineers, apothecaries, travellers and country gentlemen; rarely by university teachers. Few names—Boerhaave, Black—spring readily to mind as men whose careers were almost wholly spent in university teaching. Even Newton left Cambridge in middle age to enter public life. Scientific research could be done in universities and sometimes was, but it was not expected, and much more was done outside them.[2]

This was still the state of affairs in 1800. A century later, the universities dominated science even more than they do now, for industrial research was still in its infancy. The change did not take place in all countries or in all sciences at the same time. It originated in Germany and in chemistry, and in order to understand it we must look closely at certain aspects of the history of Europe and the history of science.

Germany before 1815

"Germany" at the end of the eighteenth century was a patchwork of independent states, ranging in size from Brandenburg-Prussia down to petty dukedoms, free cities and even free villages. Some were Catholic, some Lutheran, some mixed; all were poor; east of the Elbe, desperately poor. The former common allegiance to the Emperor was virtually forgotten, and the only unity was one of language[3] and culture. A Mecklenburger might not

179

G*

always be welcome in Bavaria, but he would be understood (with difficulty) and would feel himself to be among strangers, rather than foreigners. This unity was fostered from 1770 onwards by a remarkable flowering of literature in German and the necessary development of a written language which compromised among regional peculiarities of speech. Because there were so many frontiers, travel could be annoying, but not difficult if one's papers were in order; and there was a long tradition, especially among journeymen and students, of wandering from town to town, usually on foot. Each state, however small, was an autocracy supported by a bureaucracy which collected taxes, settled disputes, policed frontiers, and applied with more or less zeal a labyrinth of regulations, extending sometimes to the prescription of the dress to be worn by the various classes of citizen. Much of the wealth produced on the land was dissipated by the need to keep up scores of petty courts and their legions of officials and hangers-on. In Prussia (a special case) three-quarters of the revenue was spent on the army.[4]

A bureaucracy has to be literate—even for an army it is advisable—and their higher ranks need a system of higher education. Germany had nearly thirty universities,[5] compared with two in England. Most of them had originally the structure imposed on them by Melanchthon—a junior Faculty of Arts or Philosophy, which had to be passed through before entering the Faculties of Law, Theology or Medicine; but in the course of time this had usually been modified, often into two Faculties of Theology and Philosophy of approximately equal seniority. Theology trained the ministers of religion, and into Philosophy every other subject would be bundled, including Jurisprudence for the multitude of lawyers.[6] Far more lawyers and parsons were trained than there were jobs for, and the period of waiting for a suitable appointment was usually spent in teaching. The ill-paid village schoolmaster, university-trained, waiting (often in vain) for a call to a parish, or to an office of state, was a well-known figure in eighteenth-century Germany. There are many disadvantages to this method of providing mass education, but it did result in Germany, despite its poverty, becoming a highly literate nation. Education was nominally compulsory for all children in Prussia as early as 1717, though performance lagged far behind intentions; but by 1830, Prussia was "the land of schools and barracks".

The instruction in German universities, as in those of other countries at that time, was predominantly literary, in the sense of "bookish". Some science was taught—chemistry and botany, often as part of Materia Medica, mathematics, a little astronomy—but always from books rather than from the real world. The only thing that could possibly be called experimental science was the occasional anatomy demonstration to medical students.

Universities, providers of servants for the state, were financed by the state (Catholic universities usually by the Prince-Bishop). The only staff rank was that of Professor, though as time went on increasing numbers of assistant and extraordinary Professors were appointed. A student was entitled to attend the Professors' lectures on payment of a very modest fee, but the serious student who wished to graduate formally (by no means all did) usually had to pay for private tuition from a Docent (or *Dozent*), a teacher licensed by the university but receiving no salary from it.[7] A *Dozent* had to be an effective teacher to survive, and it was expected (though it did not always happen) that new Professors would be called from their ranks.

A very striking feature of the German university system is its similarity to the German guild and apprentice system. Both had their roots in the urban life of the Middle Ages, and both had survived less altered in Germany than anywhere else in eighteenth-century Europe. The guilds were corporations of master-tradesmen, analogous to the academic body of Professors. The apprentice/student, once admitted, had to serve his time—usually about four years in both cases—and eventually produce a "masterpiece"/thesis to be judged by his guild/university. In course of time, he might become a master/ Professor himself, though usually only after spending several more years as a *Geselle* (journeyman), roughly corresponding to the *Dozent* in academic life. And just as the guilds excluded the sons of men practising *unehrliche Berufe* (for example barbers, gravediggers) so the lower classes were effectively denied admission to universities by the need to know Latin.

Neither the apprentice nor the student was expected to serve his time in one place under one master. These years were his *Wanderjahre* when he went, usually on foot, from one town to another, learning his craft under different masters. Travellers in Germany seldom failed to remark on these *Wandergesellen* —apprentices, journeymen, or students—they met with so frequently on the appalling roads. One result of this practice was a levelling out of standards, both of craftsmanship and scholarship, over the whole country. More important from the present point of view is that the German university student saw himself as an apprentice to scholarship, and knew that, if he were in earnest, he would have to produce a "masterpiece" to fairly well-defined standards of excellence. This system, though not expressly designed for the encouragement of scientific research, could hardly be bettered for the purpose; and so it proved in the next century.[8]

It must be confessed that the system did not foster intellectual liveliness. Just as the guilds produced master-craftsmen indistinguishable from one another (or from the previous generation), so the universities tended to produce mediocre hard-working pedants totally without originality. Of course there were exceptions—Kant and Haller, and a few others—and the large French refugee population[9] ensured that towards the end of the eighteenth century the Enlightenment had its reflection (*die Aufklärung*) in Germany. There was also the great literary revival of this period, the poets and novelists of *Sturm und Drang*, the heady excitements of the Romantic Movement, of Pan-Germanism, of *Naturphilosophie*.[10] But until the end of the century, these stirrings had barely affected the mass of the country, and it was still true of most Germans that "their dearest wish was to go on living as they had always lived".[11]

The French Revolution and its Napoleonic sequel rudely disturbed this prospect of peaceful mediocrity. The fruits of Enlightenment, it seemed at first, were Terror and the mob; nor could they be confined within the borders of France. It is not possible, in the compass of this paper, to enter into the details of the political and military manoeuvring, the dealing and double-dealing of the years 1792–1815. The facts can be found in any standard history.[12] It must suffice to recall the astonishing destruction of the Prussian military machine at Jena and Auerstädt in 1806, and the extension of rule by Napoleon and his puppets over virtually the whole of Germany; lasting until 1813, when the Russians chased the remnants of the Grande Armée back to France, and Napoleon survived destruction at Leipzig, only to meet it at Waterloo.

Germany after 1815

Post-Napoleonic Germany was in an extraordinary state of intellectual turmoil. The old patterns of life had been broken, and—despite all the efforts of Metternich and the princelings who had regained their territories—the pieces could never again be reassembled. The French had been hated as invaders, and of course they had to go, but once gone, their stock rose remarkably.[13] They still represented the summit of European culture; and they had blown a gale of fresh air through the stuffy corridors of *Kleinstaaterei*, and opened German eyes to their own rather backward condition and to a future rich in possibilities. (Non-Prussians also remembered with quiet satisfaction how Prussian arrogance had been humbled at Jena.) One of the possibilities was represented by Britain, a country which had changed almost beyond recognition during the years when communication had been difficult. After the peace, numerous Germans came to see the new factories, the wonderful machinery worked by steam, the mushrooming industrial towns lit by gas. Could Germany ever be like this? They sent back reports in which horror and admiration were nicely blended;[14] many stayed to work, and some to become rich in Manchester, though desolate for the clear air of their homeland. The situation of the German people after 1815, and their ambiguous feelings about Britain and France, can be compared in some degree with that of the Japanese after 1853.

There was great pressure for change, increased rather than decreased by a slowly rising standard of living.[15] Politically, this pressure was stifled until it built up a head of steam that burst forth in 1848. But there was little to hinder reform in other directions, such as the universities. There were one or two models, and a new philosophy. One model was Göttingen, which had British connexions through the house of Hanover (it had been founded in 1734 by George II) and had even attracted a trickle of British students, including Thomas Young. Much of the teaching was in German rather than Latin, and the curriculum, with its emphasis on humanistic studies, especially history, was more adventurous than in most universities.[16] Another model was the *Bergakademie* at Freiberg, to which A. G. Werner had attracted students from all over the world by the liveliness of his teaching—though it was not a university, and taught only a narrow range of subjects.[17]

The philosophy to which the university reformers paid at least lip-service was that sketched out by Schelling in his *Vorlesung über die Methode des akademischen Studiums* (1803). In contrast to the largely vocational training (dreary and not always even useful) which the German universities were offering, Schelling proclaimed the function of the university to be the pursuit of Truth, in the sense of all-embracing knowledge or *Wissenschaft*—a much broader concept than "science" in English. The duty of the university teacher was not to inculcate facts (these had their proper place in encyclopaedic works of reference) but to introduce his students to the methods of seeking out the truth; the methods of research and criticism. This he could best do, not by precept, but by example, so that the teacher himself must be, not a purveyor of facts and received opinions, but a productive scholar and research worker.[18] Conversely—and here we see the ethos of the guild system still at work—the duty of the scholar must be not only to his scholarship, but to the teaching of his methods to a new generation of student-apprentices, who would in their turn go on with this unfinished—indeed, unfinishable—task. A student could begin the study of *Wissenschaft* at any point, and might end very far from where he began; there must be no constraints, either on learning or teaching (*Lern- und Lehrfreiheit*). Such was Schelling's attempt to recover a version of the medieval "universitas" for the dawning nineteenth century.

These high ideals were not likely to be put into practice by a philosopher, certainly not by a *Naturphilosoph* like Schelling. They had, of course, to suffer considerable dilution by translation into the language of everyday academic life, a task made easier by the imprecise way in which Schelling expressed himself.[19] There was opposition from the Junkers of the north, and their Catholic counterparts of the south, who did not understand these new-fangled notions; and also from the powerful student organisations, the *Bursenschaften*, who suspected that reform might be a cloak for increased political control. The rejuvenation of the German universities was not the work of one man, even of a dozen men. In the field of natural science, Alexander Humboldt, the traveller and polymath, has the reputation of being the key figure, though this is hard to establish, since his influence was exerted behind the scenes and largely undocumented. He was certainly well placed to exert influence; though he held no official position, he was well connected—his brother Wilhelm, the philologist, was for a time a Minister in the Prussian Government—and respected far outside the bounds of Prussia. The first act of reform, carried out in the darkest hour of Prussian humiliation under Napoleon, was the foundation of a new university in Berlin (1810). Successive kings of Prussia had formerly refused a university in the capital, favouring rather the Prussian Academy which was very much under the royal thumb. This was followed by another in Bonn (1818) in the Rhineland provinces taken by Prussia in the peace settlement. Both places acknowledged the ideals of Schelling; but a more important sign of the times was perhaps the appointment to the chair of chemistry in Berlin of Klaproth—who was not an academic, but an elderly representative of the mining and assaying tradition in German chemistry.

The finest fruit of the renewal of the German university system will be differently identified by people of different background and interest. Some will see it as critical historical scholarship, whose best known representative was Leopold von Ranke; others, as the creation of linguistics as a respectable branch of learning, through the labours of men like Franz Bopp. Both disciplines, though hardly "science" in the English sense, fall well within the bounds of *Wissenschaft*. But from the present point of view the significant thing was the spectacular growth of research in natural science, especially chemistry, and the birth of the university research school.

Liebig

Justus Liebig (1803–73) was a young chemistry student in Paris when he met and impressed Alexander Humboldt. He was not satisfied with German chemistry, which he had studied in Bonn and Erlangen. It was far too literary; the phlogiston theory had waned, the new French chemistry was only half understood by the professors and the resulting void was filled with a woolly mass of *Naturphilosophie*. This was not the chemistry he had fumbled with in the apothecary's shop, nor as he believed it to be in France. So to Paris he went, and eventually became a pupil of the great Gay-Lussac himself. By Humboldt's influence he was appointed assistant professor of chemistry in Giessen, at the age of barely twenty-one. A subordinate post in a small university in an unimportant state (Hesse-Cassel)—on the Humboldts' chessboard this must have seemed a minor move; for Germany and for chemistry it proved to be the most important thing they did. The sudden death of Liebig's superior in the next year (1825) saw him a full professor. In a barn-like room in a disused barracks he set about creating a school of chemical research.

It will be necessary here to digress a little, and to discuss the impact which the new chemistry was making on teaching, and on the practical skills which a chemist might be expected to have. The point of central importance was Lavoisier's concept of "element", and his list of known elements, as modified by later discoveries by Davy and others. (In the 1820s, the Atomic Theory was of lesser significance, though it was not destined to remain so for long.) Once this list and its simpler implications were accepted, an immediate practical task was offered to the chemist; to take the multitude of *things* there are in the world—rubies, carrots, Roman coins, fossil bones, horsehair, cheese, iron ore—and find out which elements they contain, and how much of each. Here was an endless series of problems to which the answers could be found only by experiment, and not by ratiocination, or by balancing one authority against another. *Analysis*—the same word as used long before to denote the resolution of bodies by heat into earth, air, fire and water, now had a new and more exact meaning. Chemists took to this enormous task with enthusiasm. It was a relief, after *Naturphilosophie*, to have something definite to do; there were no intellectual difficulties, once the list of elements had been accepted (and there was always the exciting prospect of finding a new one); in practice, the problems of analysis were hard enough to be interesting, but usually capable of solution in the end. Liebig's school was centred upon analysis and with improvements in technique (for many of which he was himself responsible) it became possible to extend the scope of analysis with confidence from the mineral kingdom to the much more complex organic realm. The new research chemist was a direct descendant of the old craftsman, the assayer.

Liebig worked in the same room, side by side with his students, not for any egalitarian reasons, but because there was no other. Formal teaching was kept to a minimum; as a young man, Liebig did not enjoy lecturing. But, as we have seen, chemistry had reached that interesting point in its development where only a small stock of theoretical ideas had to be mastered before embarking on fruitful practical work. The acquisition of technique was more essential, and took longer, than the learning of the necessary basic theory. There was (at least at the beginning) no clear-cut demarcation between teaching and research, none of the years of marching towards the front line that the young soldier of science has to do today. Practical instruction began by repeating the preparation and analysis of well-known compounds; when reliable results were obtained on these, the student went on to less well-known ones, to confirm (or contradict) the few analyses already done; then, by an insensible progression, to the preparation and analysis of new, unknown compounds, work which might well be worth publication.

Liebig was thus happily exploiting a situation rather rare in science, where a huge collection of raw and (at first sight) meaningless data is waiting to be examined, a collection which it is dimly foreseen will be the jumping-off ground for the next theoretical leap forward. In this case, analysis led to attempts at synthesis and (joined to the Atomic Theory) to the structure theory of organic chemistry, and to a new task for chemists (not yet completed) of assigning a molecular structure to everything in the world. Germany was to be pre-eminent in this too, though Liebig himself turned aside to other things. Before this happened, however, he had his triumphs in the field of analysis— the establishment of isomerism, and the discovery of "compound radicals".

But only to a small extent can we look to these scientific achievements to explain Liebig's success in attracting able students and founding a school which became famous far beyond the bounds of Hesse-Cassel.[20] At the beginning, when Giessen stood alone, the main attraction would be the novel prospect of studying chemistry with hands as well as head, and the contrast to other German schools of chemistry with their literary bias. There was undoubtedly a swing among the student generation against *Naturphilosophie*, of which Liebig was glad to take advantage. The shrewd acquisition, in 1831, of a tottering pharmaceutical journal, gave him a house journal and a vehicle of publicity which became famous as "Liebig's *Annalen*". The course was short —a hard-working student could get a doctorate in nine months if he was well prepared on entering. It was cheap, though this would only interest foreign students, since all German universities were cheap. By the late 1830s Liebig's prestige was sufficient to "bend" the entry requirements, so that any promising young man could be admitted. But the root cause of Giessen's fame must have been Liebig's personality and enthusiasm,[21] taking skilful advantage of a tide in the affairs of chemistry, and the intellectual ferment of post-Napoleonic Germany.

The "Giessen system", as I shall call it, is one of the great inventions in the organisation of science. It enables the labours of many people to be brought to a point in a way impossible before, and a mass of interlocking problems can be tackled which would daunt the lone researcher. The workers need not even be of first-rate ability, so long as they are competent; a second-rater will simply make a smaller contribution than a first-rater. But it is not, as we shall see, a recipe for automatic success, and perhaps it carries within itself the seeds of its own decline.

The Influence of Giessen

Liebig soon had imitators who, because of the continuing practice of students wandering from university to university, were fellow-pioneers rather than rivals. Indeed H. Rose, Klaproth's successor at Berlin, had started a school of analysis at about the same time as Liebig, and in the early days his reputation was equal or greater. But Rose seems to have lacked Liebig's personal magnetism, and his analyses were mostly of inorganic materials—in retrospect, a less fortunate choice. Also, since there were no facilities for practical work in the university itself, the laboratory of the Royal Prussian Academy had to be used, a situation which led to many difficulties. Rose's colleague, the brilliant Mitscherlich,[22] seems to have had no desire to create a Giessen-type department.

In 1836 Liebig's friend and collaborator Wöhler was called to Stromeyer's chair in Göttingen, a university whose willingness to innovate we have already noted. Stromeyer had already set up a teaching laboratory in Göttingen of which we know all too little. Despite the bleak political climate which followed the severance of the British connexion in 1837,[23] Wöhler was able to reorganise the chemistry school on a pattern similar to Giessen, though with a greater proportion of formal teaching. In 1838, Robert Bunsen, a *Dozent* from Göttingen, and perhaps the finest teacher of them all, founded a similar school in Marburg; in 1852 he did the same at Heidelberg. Wöhler continued to teach until 1882 and Bunsen until 1889, and many thousands of students passed through their hands. Bunsen, unmarried, totally devoted to his calling, and a very "clubbable" man, had great influence on his colleagues both at Marburg and Heidelberg; his interests were much wider than chemistry. Other Giessen-type departments were started in the 1840s by Erdmann at Leipzig (who had a rival to the *Annalen*, the *Journal für praktische Chemie*) and by the tragically short-lived Marchand at Halle. As the students of these pioneers were called to chairs, the system spread quickly during the 1850s.

Attempts to copy the Giessen style abroad were sporadic and not always successful. The most famous is the Royal College of Chemistry (1845) in London, of which Liebig was pressingly invited to become the head. He sent instead his protégé Hofmann, and the College achieved a modest success during the twenty years that Hofmann stayed there, but declined when he left. More permanent (though still modest) success can be claimed for H. E. Roscoe, a pupil of Bunsen, who went to Owens College, Manchester, in 1857. Much of the difficulty encountered in transplanting Giessen onto alien soil must be put down to the totally different university tradition. In France, Dumas evolved courses of laboratory instruction in the 1830s which had some similarity to Liebig's, though there may have been no direct influence;[24] the similarity became much stronger under his successor Wurtz, who had actually spent a year at Giessen. Unlike Germany, however, with its profusion of universities of roughly equal standing, there were at this period no provincial rivals to Paris, and appointment outside the capital was felt to be a form of exile.

Sciences other than chemistry[25] seem to have been at a less suitable stage of development for the introduction of the Giessen system, and there was a considerable time-lag. Physics, with its historical links with astronomy, has an observatory tradition which (one would think) could easily be adapted; perhaps a Liebig of physics would have set his students to measuring the physical properties of the endless number of different materials which exist or can be made. This does not sound exciting (but nor does chemical analysis, and Liebig's students found that interesting enough) yet it would surely have led to the discovery of anomalies, out of which fundamental advances might well have come. However, this did not happen. In Berlin there was Liebig's friend Poggendorff, a successful teacher with numerous students and an *Annalen der Physik* corresponding to the *Annalen der Chemie*. An even more productive teacher of physics was the rather obscure Franz Neumann, professor in Königsberg; by 1900, practically every German physicist of repute was an ex-pupil of one (or both) of these men. Though practical work was not absent from the courses at Berlin and Königsberg, there is no evidence that they ran on lines at all like Giessen. In physics, until well into the second half of the century, the typical picture was not of the productive research school, but of the isolated professor, occasionally (like Gauss) of towering importance, more often of relative insignificance. Not until 1871 was a Giessen-type physics school founded, by Helmholtz in Berlin, and he, significantly, had made his name in physiology.

The biological sciences offer great scope for comparative studies, which should make a Giessen system readily adaptable to them. The first one to develop such a system was physiology, with Johannes Müller's school in Berlin (1833–58). Müller, like Liebig, took full advantage of the reaction against *Naturphilosophie,* though he was never able to free himself entirely from its preconceptions. He believed that all scientific progress arose through *denkende Erfahrung,*[26] and he taught through laboratory work in the Giessen fashion, though direct indebtedness to Liebig cannot be proved and is perhaps unlikely. A neurotic man, prone to fits of depression, he was hardly the equal of Liebig in personality and drive, but his pupils were very numerous, including such significant figures for the next generation as Helmholtz (already mentioned) and Du Bois-Reymond. The Giessen tradition is much more obvious in Carl Ludwig, who created experiment-based departments of increasing size and importance at Zurich (1849), Vienna (1855) and Leipzig (1865). Ludwig had been student, *Dozent,* and extraordinary professor at Marburg, where he came much under the influence of Bunsen, and through him, of Liebig.[27] He was a fertile inventor of simple but versatile instruments, with which even his less gifted students could make an endless series of useful measurements—and since he was of an unusually selfless disposition, the results were often published under the student's name alone.[28] Botany and zoology, however, appear to have been much slower to adopt the Giessen system.

The reason for this time-lag between the organisation of research in chemistry and in other sciences may lie in the relative absence of career opportunities. The young German went through a university course in the expectation of having to earn his living; he was not a gentleman, as he was in England and in most other countries. He expected to do a job, and there have always been jobs for chemists: in pharmacy, in mining and assaying, in the textile trade, increasingly in chemical manufacture. The same is true, to a slightly less degree, of physiologists, whose work is the basis of the important "industry" of medicine. There were, until the last quarter of the century, few such opportunities in physics, botany or zoology. When, later, such opportunities arose—in physics with the rise of the electrical industry, in economic botany and zoology (especially colonial), and above all in the "new" science of microbiology with its numerous medical and industrial applications— Giessen-type schools in these sciences quickly sprang up to meet the demand.

Though the Giessen system did not readily suffer transplantation outside Germany, from 1835 onwards it attracted increasing numbers of young foreigners into Germany. People came from Britain, Holland, Scandinavia, Italy, Russia, the United States, even (though not so many) from France; some even came from unexpected places like Portugal or Mexico. They came because they could not get such an education at home, because it was cheap— fees were low, there was no residential system, no social position to keep up, and German provincial towns offered few expensive distractions; they got a Ph.D., and the training they were given offered an entrée into an interesting and (perhaps) remunerative career. Apart from all this, Germany must have been a very pleasant place to live in at this period. Liberal ideas were abroad, prosperity was beginning to touch even the poorest, industrialisation of an idyllic countryside had hardly begun.[29] All was not as well as it seemed, as the events of 1848 were to show. But British students, at any rate, seem to have thoroughly enjoyed their time in Germany, and often returned with a German bride. By 1900 there was hardly a chemist of any standing in Britain who had not a German Ph.D.—a degree still not obtainable in his own country.

It would be pleasant to take leave of Liebig at the summit of his greatness, working in his shirt-sleeves in his soot-grimed makeshift laboratory, among a motley crowd of students from all nations. But history moves on, even past happy endings, and the historian has to chronicle the change from Liebig the chemist and teacher to the less attractive figure of Liebig the Public Man. Nor did the ideals of Giessen survive untarnished, though they are still quite recognisable even today. Some of the tarnish was peculiarly German, some came from weaknesses there from the beginning. In Germany, chauvinism and militarism spread slowly over the universities, first the Prussian, then the non-Prussian ones, and foreigners gradually found themselves less welcome.[30] Bismarck's *Kulturkampf*, his harassment of the Social-Democrats, his aggressive Prussianism, made many Germans uneasy to the point of neurosis and strange philosophical and pseudo-scientific movements began to flourish.[31] Outside working hours in the universities, the duelling clubs elbowed out the rowdy sing-song in the *Bierstube*. Nor would the professor be found in the *Bierstube* as Bunsen so often was; the distance between professor and student widened to an impassable gulf as departments grew larger. "Education through research" can, unless great care is taken (possible only in a small and friendly department) result in deep but narrow knowledge, and the blinkered specialist, the semi-literate Ph.D., was already a problem figure in Germany by the 1880s.

This last problem was by no means exclusive to Germany, but arose wherever there were large research schools. The influence of the guild system waned and was supplemented by the influence of the factory system, which saw the student less as an apprentice than as a "pair of hands" working for the greater glory of his supervisor, the department as a conveyor belt for the production of Ph.Ds, the publication of scientific papers as a sort of dividend. Eventually the professor, and even the junior members of his staff become managers, concerned with administration and output, rather than discovery. The lone wolf, pursuing his own line independently of the big departments, is discouraged in a variety of non-violent but effective ways. The vast bulk of published scientific work nowadays is actually (whatever names may appear at the head of the paper) the work of junior apprentices, not of master-craftsmen—a fact which should give us pause for thought more often than it seems to do.

All these faults which loom so large with us now can be seen in embryo by the historian who studies Liebig and his school in Giessen. Will they bring down modern university science, as the barren pedantry of the later schoolmen brought down the medieval university? And will the science of the future be pursued in some quite different way?

References

1. For instance, in the Journal of the Chemical Society for 1962—an average scientific journal in an unremarkable year—only 9.3 per cent of the contributions came from industry (including some of joint industry-university authorship). Even at the "applied" end of the same subject (*J. Appl. Chem.*, 1964) 37.5 per cent of the papers came from universities.
2. The Italian universities are a partial exception to this generalisation. Spallanzani, Galvani, Volta and many lesser men were university teachers. But in general the relationship was like that existing today between the universities and creative literature.
3. The Prussian language died out before the eighteenth century, and the small Wendish-speaking area in Saxony has never been important. On the other hand, the parts of Poland seized by Prussia in the partitions were never assimilated linguistically, except in border areas.
4. For a general survey of life in Germany at this period, see W. H. Bruford (1935), *Germany in the Eighteenth Century* (Cambridge).
5. This means the territory of present-day Germany (East and West) together with Königsberg, now in the USSR. There were also four universities in present-day Austria, and a number of others under varying degrees of German influence: Basel, Strasbourg, Prague, Cracow, Posen, Dorpat (founded 1802), Zurich (founded 1833).
6. There is a considerable literature on the history and organisation of the German universities. The standard work is still F. Paulsen (1921), *Geschichte des gelehrten Unterrichts* ... (1885), of which I have used the 3rd edition (Berlin and Leipzig), especially vol. 2 (*Der gelehrte Unterricht im Zeichen des Neuhumanismus, 1740–1892*).

Paulsen, however, is interested in the classical tradition, and pays no special attention to science.

7. There are resemblances to the "crammers" of Oxford and Cambridge though these were not licensed or recognised by the university. See W. J. Reader (1966), *Professional Men* (London).

8. In contrast, the only serious purpose of most English undergraduates in the eighteenth century would be to form acquaintanceships which might be useful in the years to come.

9. The Great Elector answered the revocation of the Edict of Nantes (1685) with his own "Edict of Potsdam" offering asylum in Brandenburg-Prussia to all Huguenots. A century later, according to some estimates, a quarter of the population of Berlin was of French descent.

10. On these subjects (except the last) see R. Pascal (1953), *The German Sturm und Drang* (Manchester). On *Naturphilosophie*, see the chapter by D. Knight; also B. Gower (1973), *Studies Hist. Phil. Sci.*, **iii**, 301–56.

11. Mme. de Staël (1814), *De l'Allemagne* (Paris)—but referring to an earlier period): "Le plus vif désir des habitants de cette contrée paisible et féconde, c'est de continuer à exister comme ils existent".

12. For instance, G. P. Gooch (1920), *Germany and the French Revolution* (London).

13. Cf. Goethe's observation: "Ich hasste die Franzosen nicht, wiewol ich Gott dankte, als wir sie los waren. Wie hätte auch ich, dem nur Cultur und Barbarei Dinge von Bedeutung sind, eine Nation hassen können, die zu den cultivirtesten der Erde gehört, und der ich einen so grossen Theil meiner eigenen Bildung verdankte!" (J. P. Eckermann, *Gespräche mit Goethe*, 14 March 1830).

14. Compare the bucolic England described by Pastor Moritz (C. P. Moritz, *Reisen eines Deutschen in England im Jahre 1782*) with the German and Swiss-German reports quoted by W. O. Henderson, *Industrial Britain under the Regency* (London, 1968).

15. Agricultural improvements (crop rotation, introduction of potatoes, etc.) came late to Germany, but their impact was all the greater. There was also, after 1815, a great reduction in the area of land left deliberately uncultivated so that the nobility might hunt.

16. On Göttingen, see Paulsen, ref. 6.

17. Cf. Cuvier's éloge of Werner (1818): "C'est ainsi qu'en peu d'années la petite école de Freyberg, destinée seulement, dans le principe, a former quelques mineurs pour la Saxe, renouvela le spectacle des premières universités du moyen âge; qu'il y accourut des élèves de tous les pays où il existe quelque civilisation; et que, dans les contrées les plus eloignées, l'on vit des hommes déjà sur l'âge, des savans déjà renommés, se hâter d'étudier la langue allemande uniquement pour se mettre en état d'aller entendre le grand oracle de la géologie". See also *Bergakademie Freiberg: Festschrift 1765–1965* (Freiberg, 1965); D. M. Farrar (1971), *The Royal Hungarian Mining Academy, Schemnitz: some aspects of technical education in the 18th century* (M.Sc. Thesis, Manchester), chapters 5 and 6.

18. By "research" Schelling did not mean experimental work in a laboratory, though after consideration he might well have approved of it.

19. A useful exercise would be to take the *Vorlesung* and attempt to turn it into concrete recommendations likely to be accepted (or even understood) by an Academic Development Committee.

20. For more on Liebig's research school (and comparison with Thomson's less successful attempt in Edinburgh) see J. B. Morrell (1972), *Ambix*, **xix**, 1–46.

21. An aspect of Liebig's enthusiasm that has never, as far as I know, been stressed

before, is his encouragement of former students to continue doing research long after they had returned home. This is shown, for example, by two letters from the Manchester chemist Edward Schunck to Liebig (8 August 1844 and 26 October 1852) in the *Bayerische Staatsbibliothek*.

22. Mitscherlich is a splendid example of *Lern- und Lehrfreiheit*. He took his doctorate in Old Persian, made his name in crystallography (Law of Isomorphism), went on to do good work in organic chemistry, and spent his last years studying volcanoes.

23. When Victoria came to the British throne in 1837, she could not become Queen of Hanover because of a different law of succession. The Duke of Cumberland (one of the "wicked uncles") became king, and before long he deprived seven Göttingen professors of their chairs for political reasons.

24. Dumas' models may equally well have been the practical classes of the École Polytechnique and the collaborative research (on taxonomy, etc.) carried out at the Muséum d'Histoire Naturelle.

25. Mathematics is not an experimental science, and the situation does not arise.

26. Roughly "thought-out experience"—experience evaluated within the framework of a philosophy.

27. "Sich selbst mit präciser physikalische Methodik vertraut zu machen, hatte [Ludwig] in Laboratorium des damals in Marburg weilenden Robert Bunsen Gelegenheit gehabt. Als Vorbild endlich für die Wirksamkeit eines wissenschaftlichen Lehrers, eines Organisations der Forscherarbeit im grössten Stil mochte ihm Liebig im nahen Giessen gelten; wie aufmerksam Ludwig dessen geistiges Wesen und Treiben verfolgt hat, beweist der feine Nachruf, den er ihm—anonym —in der Wochenschrift "Im neuen Reich" gewidmet" (A. Dove in *Allgemeine deutsche Biographie*, **55**, 896). "Family trees" of pupils, both of Müller and Ludwig, are given in K. E. Rothschuh (1973), *History of Physiology* (Huntington, N.Y.).

28. "Der betreffende Russe oder Holländer stand dabei, hielt etwa den Schwamm oder das Handtuch, wusste kaum, was vorging, am allerwenigsten den Gedankengang der Sache, liess sich einige Zahlen in sein Notizbuch dictiren und war nachher erstaunt, eine wunderschöne Arbeit unter seinem alleinigen Namen gedruckt zu sehen" (F. Miescher, quoted by Dove, ref. 27).

29. In Germany this is sometimes called the "Biedermeyer" period, after a popular novelist ("Dickensian" might be a rough English equivalent, but the Dickens of *Pickwick*, not of *Oliver Twist*). For a contemporary English view, see W. C. Perry (1846), *German university education* (2nd edn. London); and in lighter vein, W. Howitt (1841), *The student life of Germany* (London). Matthew Arnold's eulogy in *Schools and universities on the Continent* (London, 1868) is well known; it concludes "It is in science that we have most need to borrow from the German universities. The French universities have no liberty, and the English universities have no science; the German universities have both."

30. See, for example, E. Kennaway (1952), *Ann. Sci.*, **viii**, 393–7, for an account of his time in Kossel's laboratory in Heidelberg; K. Mendelssohn (1973), *The world of Walther Nernst; the rise and fall of German science* (London); J. Ben-David (1968–9), *Minerva*, **vii**, 1–35; A. Flexner (1930), *Universities: American, English, German* (Oxford). I am also indebted to my former headmaster, the late John Hughes Davies, Ph.D.(Leipzig)—one of the last generation of British chemists to study in Germany before 1914—for his reminiscences, to which I now wish I had paid more attention.

31. See D. Gasman (1971), *The scientific origins of National Socialism* (London). Gasman marshals impressive evidence for Ernst Haeckel, a pupil of Müller, as a philosophical godfather to the Nazi movement.

Name Index

Place Index

Africa, 173
America, Latin, 174
Amsterdam, 92
Antarctica, 173
Australia, 173
Austria, 162
Avernia, 41

Bavaria, 162
Berlin, 174, 184, 188
Bologna, 40, 42, 43, 51
Bordeaux, 5
Breda, 92

Cambridge, 6, 10, 82, 93, 123

Delft, 94
Deventer, 92
Dijon, 5
Dordrecht, 92
Dublin, 75

Edinburgh, 93, 115, 118, 120
England, 5, 6, 57–87

Ferrara, 40, 49, 50
France, 5, 10, 127–159
Franeker, 92, 96
Freiburg, 6, 174, 191

Germany, 5, 9, 161–192
Giessen, 184–187
Glasgow, 120
Göttingen, 172, 182, 183, 187
Groningen, 92
Halle, 187
Hamburg, 59, 73
Hanover, 161
Harderwijk, 92, 94, 97
Heidelberg, 187
Hertogenbosch, 92
Hesse-Cassel, 184–186
Holland, *see* Netherlands

Ireland, 6
Italy, 5, 15–56

Königsberg, 188

Leipzig, 187, 188
Leyden, 91–93, 97
London, 75
Lyons, 5, 42, 150

Maastricht, 92
Manchester, 6, 187
Marburg, 187, 188
Messina, 41, 44, 50
Montpellier, 5

Naples, 43, 51
Netherlands, 6, 89–109
Niger, 173
Nijmegen, 92
Nuremberg, 42

Oxford, 36, 75, 82

Padua, 20, 36, 37, 40, 41, 44–47, 49, 51
Paris, 4, 5, 36, 73, 95, 123
Pisa, 40, 41, 45–47, 50
Prussia, 162, 180, 183, 184

Ravenna, 41
Rome, 40, 41, 50, 79, 80
Rostock, 60
Russia, 161, 174

Schemnitz, 6
Scotland, 5, 111–126
Siberia, 174
Spain, 19

Toulouse, 80
Tuscany, 15, 16

Utrecht, 91, 92, 97

Venice, 3, 16
Vienna, 188

Zurich, 188

Subject Index